Electro-Chemo-Mechanical Properties of Solid Electrode Surfaces

Masahiro Seo

Electro-Chemo-Mechanical Properties of Solid Electrode Surfaces

Springer

Masahiro Seo
Emeritus Professor, Faculty of Engineering
Hokkaido University
Sapporo, Japan

ISBN 978-981-15-7276-0 ISBN 978-981-15-7277-7 (eBook)
https://doi.org/10.1007/978-981-15-7277-7

This Springer imprint is published by the registered company Springer Nature Singapore Pte Ltd.
The registered company address is: 152 Beach Road, #21-01/04 Gateway East, Singapore 189721, Singapore

Preface

Surface stress is the mechanical work unique for a solid surface which is elastically deformed due to low mobility of the component atoms. When the potential is applied to a solid metal electrode in electrolyte solution, the surface stress of the solid electrode changes depending on the applied potential due to the electrochemical reaction taking place on the electrode such as electro-sorption or electro-deposition. Reversely, when a solid metal electrode is elastically strained, the potential of the electrode varies depending on the degree of the surface strain. In addition, when a surface oxide film (anodic oxide film) is formed and grown on a solid metal electrode during anodic oxidation, stress is generated in the film and the film stress varies depending on the conditions of anodic oxidation. The properties of solid electrode surfaces associated with the correlation between electrochemical and mechanical phenomena are named "electro-chemo-mechanical properties".

This book deals with the electro-chemo-mechanical properties of solid electrode surfaces that are useful for a deep insight into the interfacial phenomena such as electro-catalytic reaction, surface film formation and corrosion. This text is aimed at graduates and senior undergraduates studying interfacial electrochemistry and materials science or those beginning research work in the fields and related areas.

In Chap. 1, surface thermodynamics of a solid electrode is derived as fundamentals for understanding of electro-chemo-mechanical properties. This chapter is based on "Advanced Course of Interfacial Electrochemistry" that I taught for graduate students at Faculty of Engineering, Hokkaido University. Surface stress is equal to surface tension for a liquid metal (e.g., mercury) electrode which is plastically deformed due to high mobility of the component atoms, while surface stress is not equal to surface tension for a solid metal electrode. The thermodynamic parameters theoretically derived are compared with those obtained experimentally to understand the difference between surface stress and surface tension for a solid metal electrode. Measurement of changes in surface stress of a solid electrode is indispensable for investigating the electro-chemo-mechanical properties of the electrode surfaces. The principles and features of typical methods for measuring changes in surface stress such as piezoelectric detection and cantilever bending are described in Chap. 2. In addition, new techniques of electrochemical response to elastic deformation and dilatometry for stress measurement of a nano-porous electrode are also included in Chap. 2.

In Chap. 3, potential-induced or adsorbate-induced changes in surface stress associated with surface reconstruction or adsorption of electrolyte species are explained as concrete examples for electro-chemo-mechanical properties of solid electrode surfaces. Two characteristic parameters of the surface stress-surface charge density coefficient and of the potential-surface elastic strain coefficient are experimentally compared to confirm that both parameters are theoretically equivalent each other. Underpotential deposition (UPD) of metal atoms on a foreign metal electrode in electrolyte solutions containing the corresponding metal ions has been extensively investigated in the field of electro-catalysis from the viewpoint of the enhancement of catalytic activity due to UPD. In Chap. 4, the electro-chemo-mechanical properties of the typical UPD layers on Au (111) electrode are discussed from the changes in surface stress measured by a cantilever bending and from the structural changes of the UPD layers observed by various in-situ analytical tools. There have been controversial arguments in the electrochemistry community for surface thermodynamics associated with surface stress and surface tension of a solid electrode. The controversial discussions still continue up to now. In Chap. 5, the controversial arguments made so far for surface thermodynamics of a solid electrode are reviewed to find a clue for solving the thermodynamic issues.

Stresses generated in anodic oxide films on a solid metal electrode are influenced by many factors associated with the film formation and growth mechanism. In Chap. 6, the main factors influencing the film stress are separately explained and the contributions of respective factors are discussed for the typical experimental results. Nano-indentation technique is a powerful tool to measure the mechanical properties such as hardness, elastic modulus, and yielding strength in a nanometer's range (i.e., nano-mechanical properties) of solid surfaces. In Chap. 7, the fundamentals of nano-indentation technique are explained and the nano-mechanical properties of bare metal surfaces, bulk metal oxides and anodic oxide films on metals are described as typical examples of the application results. Furthermore, the results obtained by electrochemical nano-indentation on passive metal electrodes in electrolyte solution are discussed in Chap. 7.

I would like to acknowledge the colleagues of Corrosion Research Group at Hokkaido University who gave me a strong motivation for the preparation of this book. I wish to thank postdoctoral and graduate students studied in my laboratory for their contribution to this book. Finally, I am grateful to my wife, Hiroko, for her constant support.

Sapporo, Japan
June 2020

Masahiro Seo

Contents

Chapter 1
Surface Thermodynamics of Solid Electrode

Abstract Surface thermodynamics is fundamentals for understanding of the electro-chemo-mechanical properties of solid electrode surfaces. The concept of surface phase (or interphase) is explained, and the general equations of surface thermodynamics are derived. Surface stress is the most important thermodynamic parameter to characterize a solid electrode, since the changes in surface stress are directly associated with the electro-chemo-mechanical properties. The relationship between surface stress and surface tension for a solid electrode is discussed to distinguish from surface tension which characterizes a liquid electrode such as mercury electrode. Furthermore, the electrocapillary curve (surface stress vs. potential or surface tension vs. potential) for the electrified interface of a solid electrode is derived and compared with that (surface tension vs. potential) derived for the electrified interface of a liquid electrode. The difference between electrocapillary curves derived for the electrified interfaces of solid and liquid electrodes is confirmed from the results obtained experimentally for the electrified interfaces of gold and mercury electrodes.

Keywords Surface thermodynamics · Solid electrode · Surface stress · Surface tension · Electrocapillary curve

1.1 Introduction

Surface thermodynamics is fundamentals for understanding of the electro-chemo-mechanical properties of solid electrode surfaces. This chapter deals with surface thermodynamics of a solid electrode. In the interface region of two phases, the properties vary between those of bulk phase. Surface thermodynamics treats the interface region of two phases as one phase, i.e., surface phase (or interphase). At first, the explanation for the concept of a surface phase (or interphase) has to precede the derivation of the general equations of surface thermodynamics. Surface stress is the most important thermodynamic parameter to characterize a solid electrode since the changes in surface stress are directly associated with the electro-chemo-mechanical properties. The derivation of the relationship between surface stress and surface tension for a solid electrode is indispensable for distinguishing a solid electrode from a liquid electrode such as mercury electrode. The electrocapillary curve

M. Seo, *Electro-Chemo-Mechanical Properties of Solid Electrode Surfaces*,
https://doi.org/10.1007/978-981-15-7277-7_1

(surface stress vs. potential or surface tension vs. potential) derived for the electrified interface of a solid electrode is different from that (surface tension vs. potential) derived for the electrified interface of a liquid electrode. The comparison between the electrocapillary curves obtained experimentally for the electrified interfaces of gold (solid) and mercury (liquid) electrodes is necessitated to confirm the difference between electrocapillary curves derived thermodynamically for the electrified interfaces of solid and liquid electrodes.

It is noticed that the term "solid electrode" used in this book represents simply the case where a solid electron-conductor such as metal is immersed in a liquid ionic-conductor (i.e., electrolyte solution). The term "surface" is usually used for the termination of a solid or liquid phase faced to vacuum. On the other hand, the boundary between two phases (e.g., solid and liquid) is named "interface." Nevertheless, it is reminded that the term "electrode surface" in this book is used for the electrode/solution interface unless otherwise noticed. We start this chapter from the definition of a surface phase.

1.2 Definition of Surface Phase

There are two approaches to define thermodynamically the interface region as a surface phase [1, 2]. Figure 1.1 is useful to explain the definition of a surface phase [3]. Figure 1.1a shows schematically the variation of extensive property X as a function of distance z across the interface region between bulk α (e.g., solid) and β (e.g., liquid) phases contacted each other. In Fig. 1.1a, it is assumed that the x-y plane of a rectangular prism consisting of α and β phases is flat and has a homogeneous property. One approach for the definition of a surface phase has been achieved by Gibbs [1] who replaces a real interface region by a dividing surface as a two-dimensional surface phase. As depicted in Fig. 1.1b, the two-dimensional surface phase (σ phase) is defined by the requirement of surface discontinuity that the extensive property under consideration should maintain a uniform value in each bulk phase up to the dividing surface (perpendicular line: DS) and the net value of extensive property corresponding to the areas with plus sign (β phase) and minus sign (α phase) is fixed as a surface excess of the extensive property at the dividing surface. The net value of a certain property at the dividing surface can be made zero by choosing the position of the dividing surface to equate the area of plus sign to that of minus sign. However, the net values of the other properties are not zero at the dividing surface chosen above. The demerit of this approach is that the net value of the extensive property at the dividing surface depends on the position of the dividing surface. Nevertheless, as explained later in the section of surface excess quantities, this issue is solved by adoption of the surface excess quantities of other components relative to a specified component (i.e., relative surface excess quantity) since the relative surface excess quantity is independent of the position of the dividing surface.

An alternative approach has been made by Guggenheim [2] in which the surface phase is defined as a three-dimensional phase (π phase) with a thickness of τ as

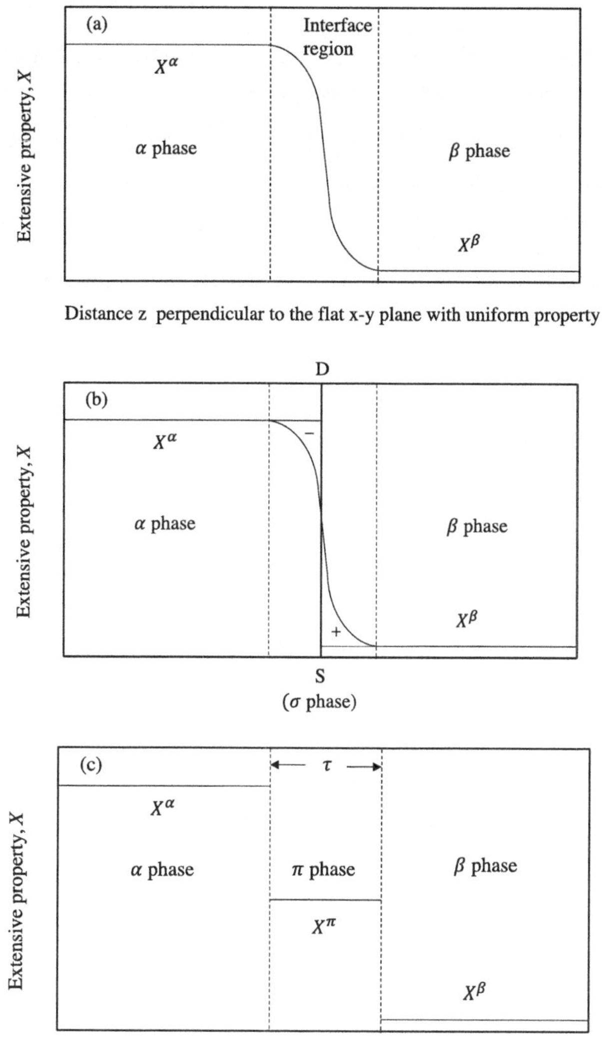

Fig. 1.1 Definition of a surface phase [3]: **a** schematic variation of extensive property X as a function of distance z across the interface region between bulk α (e.g., solid) and β (e.g., liquid) phases contacted each other, **b** two-dimensional surface phase (σ phase) defined by the requirement of surface discontinuity that the extensive property under consideration should maintain a uniform value in each bulk phase up to the dividing surface (perpendicular line: DS) and the net value of extensive property corresponding to the areas with plus sign (β phase) and minus sign (α phase) is fixed as a surface excess of the extensive property at the dividing surface [1], and **c** surface phase defined as a three-dimensional phase (π phase) with a thickness of τ [2]. Modified from [3], Copyright 2012, Springer, Berlin Heidelberg, with permission from Springer Nature

depicted in Fig. 1.1c. In the latter approach, two dividing surfaces at each boundary must be taken into consideration in the sense of the Gibbs approach. Further disadvantage is that terms dependent on surface volume are present in the equations, and it encounters difficulties in assigning values to these terms. As shown in Fig. 1.1a, the extensive property in the real interface region should vary from the border of α phase to that of β phase. On the other hand, in the Guggenheim approach, an averaged value of the extensive property X^{π} maintains uniform in the surface phase with a thickness of τ. Let the area of the surface or interface in the entire system be denoted by A. The total volume of the system V is the sum of the volumes of α and β phases and the volume of surface phase (π phase):

$$V = V^{\alpha} + V^{\beta} + V^{\pi}. \tag{1.1}$$

The volume V^{π} of the surface phase is given by

$$V^{\pi} = \tau A. \tag{1.2}$$

In the case of the Gibbs model, the surface phase has no volume, that is,

$$V = V^{\alpha} + V^{\beta}. \tag{1.3}$$

1.3 Surface Excess Quantities

In the Gibbs model, the surface excess quantity X^{σ} is given by

$$X^{\sigma} = X - X^{\alpha} - X^{\beta}, \tag{1.4}$$

where X is the sum of extensive quantity in the entire system, and X^{α} or X^{β} is the extensive quantity of α or β phase. As examples, surface excess entropy S^{σ}, internal energy U^{σ}, Helmholtz free energy F^{σ}, and Gibbs free energy G^{σ} can be written, respectively, as follows:

$$S^{\sigma} = S - S^{\alpha} - S^{\beta}. \tag{1.5}$$

$$U^{\sigma} = U - U^{\alpha} - U^{\beta}. \tag{1.6}$$

$$F^{\sigma} = F - F^{\alpha} - F^{\beta}. \tag{1.7}$$

$$G^{\sigma} = G - G^{\alpha} - G^{\beta}. \tag{1.8}$$

In addition, a quantity which is of interest in connection with adsorption and segregation phenomena is the excess amount of material or excess number of moles with relation to the interface. The surface excess number of moles n_i^σ of component i is defined as

$$\begin{aligned} n_i^\sigma &= n_i - \left(n_i^\alpha + n_i^\beta\right) \\ &= n_i - \left(c_i^\alpha V^\alpha + c_i^\beta V^\beta\right) \\ &= n_i - c_i^\alpha V + \left(c_i^\alpha - c_i^\beta\right) V^\beta, \end{aligned} \tag{1.9}$$

where n_i is the total number of moles of component i in the entire system, n_i^α and n_i^β the numbers of moles of component i, c_i^α and c_i^β the concentrations (moles per unit volume) of component i in the homogenous α and β phases, respectively, and V is the sum of V^α and V^β. Figure 1.2 shows schematically the concentration profile of component i near the interface between α and β phases. The net area marked with plus sign (+) in the concentration profile of component i due to the supposed discontinuity of c_i^α and c_i^β at the dividing surface DS in Fig. 1.2 corresponds to n_i^σ. The surface excess number of moles n_1^σ for component 1 is also defined as

$$n_1^\sigma = n_1 - c_1^\alpha V + \left(c_1^\alpha - c_1^\beta\right) V^\beta. \tag{1.10}$$

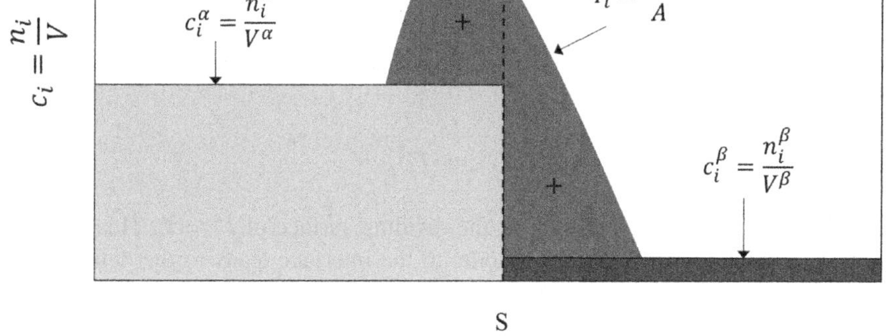

Fig. 1.2 Schematic concentration profile of component i near the interface between α and β phases. The net area marked with plus sign (+) in the concentration profile of component i due to the supposed discontinuity of c_i^α and c_i^β at the dividing surface DS in Fig. 1.2 corresponds to n_i^σ. The quantity of n_i^σ divided by the dividing surface area A; i.e., $\Gamma_i = \frac{n_i^\sigma}{A}$ is named "surface excess of component i."

In Eqs. (1.9) and (1.10), n_i, n_1, c_i^{α}, c_i^{β}, c_1^{α}, c_1^{β}, and V are dependent only of the state of the system and are independent of the location of the dividing surface. In contrast, n_i^{σ}, n_1^{σ}, and V^{β} depend on the location of the dividing surface. The following relationship between n_i^{σ} and n_1^{σ} [4] is derived by eliminating V^{β} between Eqs. (1.9) and (1.10):

$$n_i^{\sigma} - n_1^{\sigma}\left(\frac{c_i^{\alpha} - c_i^{\beta}}{c_1^{\alpha} - c_1^{\beta}}\right) = \left(n_i - c_i^{\alpha}V\right) - \left(n_1 - c_1^{\alpha}V\right)\left(\frac{c_i^{\alpha} - c_i^{\beta}}{c_1^{\alpha} - c_1^{\beta}}\right). \tag{1.11}$$

The right-hand side of Eq. (1.11) consists only of quantities independent of the location of the dividing surface. Consequently, the value of the left-hand side is invariant, irrespective of the location of the dividing surface. The quantity of n_i^{σ} divided by the dividing surface area A is represented by

$$\Gamma_i = \frac{n_i^{\sigma}}{A}, \tag{1.12}$$

where Γ_i is named "surface excess of component i." The quantity of the left-hand side divided by A in Eq. (1.11) is named "relative surface excess of component i with respect to component 1," and is denoted by $\Gamma_{i,1}$:

$$\Gamma_{i,1} = \Gamma_i - \Gamma_1\left(\frac{c_i^{\alpha} - c_i^{\beta}}{c_1^{\alpha} - c_1^{\beta}}\right). \tag{1.13}$$

Since $\Gamma_{i,1}$ is independent of the location of the dividing surface, even though both Γ_i and Γ_1 depend on the location, $\Gamma_{i,1}$ for any arbitrary location z is given by

$$\Gamma_{i,1} = (\Gamma_i)_z - (\Gamma_1)_z\left(\frac{c_i^{\alpha} - c_i^{\beta}}{c_1^{\alpha} - c_1^{\beta}}\right). \tag{1.14}$$

In the case where the location of the dividing surface is chosen at $\Gamma_1 = 0$, $\Gamma_{i,1}$ reduces to

$$\Gamma_{i,1} = (\Gamma_i)_{z0}. \tag{1.15}$$

Therefore, $\Gamma_{i,1}$ is equivalent to Γ_i at the dividing surface of $\Gamma_1 = 0$. This explains why the application of the Gibbs model to the interface leads to the definition of quantities which has a direct experimental significance.

Although Eq. (1.9) contains the terms of concentration and volume, n_i^{σ} or Γ_i can be expressed in terms of mole fraction:

$$n_i^{\sigma} = n_i - n^{\alpha}x_i^{\alpha} - n^{\beta}x_i^{\beta}, \tag{1.16}$$

or

$$\Gamma_i = \frac{1}{A}\left(n_i - n^\alpha x_i^\alpha - n^\beta x_i^\beta\right), \tag{1.17}$$

where n^α and n^β are the total number of moles of all components in α and β phases, while x_i^α and x_i^β are the mole fraction of component i in α and β phases. The use of mole fraction is convenient since one of components is not independent variable from the relationship of $\sum_{i=1}^{i} x_i = 1$. Nonetheless, the formulation of the relative surface excess in terms of mole fraction is not simple since the Gibbs–Duhem relationship for the α and β phases is needed to accomplish this purpose. We express the relative surface excess in terms of mole fraction in Sect. 1.5 of this chapter (see Eq. (1.49)).

1.4 Surface Plastic and Elastic Strains

In the case where the area of the interface (surface area) between two liquid phases or between liquid and gas phases (filled with vapor of liquid) is changed (stretched or shrunk), there is no barrier to prevent atoms (or molecules) from entering or leaving the interface region. The easy migration of atoms leads to a new state of equilibrium, in which each surface atom occupies the same area per atom in the original undistorted state but the number of the atoms in the interface region has changed. The above distortion of the surface is named "surface plastic strain." The total surface area A can be represented by the product of the number N of surface atoms and the area per surface atom a:

$$A = Na. \tag{1.18}$$

The differential of A is given by

$$dA = adN + Nda. \tag{1.19}$$

In the case of surface plastic strain ($da = 0$), dA is given by

$$dA = adN. \tag{1.20}$$

The plastic strain can also arise in a solid near at its melting point. In contrast, for a solid far from its melting point, a change of surface area encounters the difficulty in migration of atoms from and to the surface, resulting from internal constraint due to the presence of long-range order. As a result, the number of atoms of the solid remains constant while the area occupied by each atom differs from that in the original undistorted state. This type of surface distortion is named "surface elastic strain." In the case of surface elastic strain ($dN = 0$), dA is given by

$$dA = Nda. \tag{1.21}$$

In the case where a solid is subjected to partly plastic and elastic deformation, the total strain may be divided into two contributions of plastic and elastic strains as represented by

$$\frac{dA}{A} = \frac{adN + Nda}{Na} = \frac{dN}{N} + \frac{da}{a}, \tag{1.22}$$

where $\frac{dN}{N}$ and $\frac{da}{a}$ are the contributions of plastic and elastic strains, respectively.

1.5 Major Parameters of Surface Thermodynamics

In the frame of the Gibbs model, the excess internal energy U^σ is a function of excess entropy S^σ, surface area A, and excess moles of components $n_1^\sigma, \ldots, n_i^\sigma$:

$$U^\sigma = U^\sigma\left(S^\sigma, A, n_1^\sigma, \ldots, n_i^\sigma\right). \tag{1.23}$$

The complete differential of U^σ is

$$dU^\sigma = \left(\frac{\partial U^\sigma}{\partial S^\sigma}\right)_{A, n_1^\sigma \ldots n_i^\sigma} dS^\sigma + \left(\frac{\partial U^\sigma}{\partial A}\right)_{S^\sigma, n_1^\sigma \ldots n_i^\sigma} dA + \sum_i \left(\frac{\partial U^\sigma}{\partial n_i^\sigma}\right)_{S^\sigma, A, n_{j\neq i}^\sigma} dn_i^\sigma. \tag{1.24}$$

Furthermore, from the thermodynamic relationships of $\left(\frac{\partial U^\sigma}{\partial S^\sigma}\right)_{A, n_1^\sigma \ldots n_i^\sigma} = T^\sigma$, $\left(\frac{\partial U^\sigma}{\partial A}\right)_{S^\sigma, n_1^\sigma \ldots n_i^\sigma} = \gamma$, and $\left(\frac{\partial U^\sigma}{\partial n_i^\sigma}\right)_{S^\sigma, A, n_{j\neq i}^\sigma} = \mu_i^\sigma$, Eq. (1.24) can be described as follows:

$$dU^\sigma = T^\sigma dS^\sigma + \gamma\, dA + \sum_i \mu_i^\sigma dn_i^\sigma, \tag{1.25}$$

where γ is the interfacial intensive parameter conjugate to the extensive parameter A. The second term γdA in the right-hand side of Eq. (1.25) corresponds to the reversible work for creating new surface by dA under plastic deformation. In this book, γ is named "surface tension" in order to distinguish from "surface stress" associated with the reversible work for changing surface area by dA under elastic deformation. We explain the relationship between surface tension and surface stress in Sect. 1.6 of this chapter.

According to Euler's theorem since U^σ is a homogeneous function of the first order with respect to all variable:

$$U^\sigma = T^\sigma S^\sigma + \gamma A + \sum_i \mu_i^\sigma n_i^\sigma. \tag{1.26}$$

If α, σ, and β phases are in thermodynamic equilibrium, $T^\sigma = T^\alpha = T^\beta = T$ and $\mu_i^\sigma = \mu_i^\alpha = \mu_i^\beta = \mu_i$ hold so that superscripts of T and μ_i are not needed. In the bulk α and β phases, the following relationships hold:

$$U^\alpha = TS^\sigma - PV^\alpha + \sum_i \mu_i n_i^\alpha, \tag{1.27}$$

and

$$U^\beta = TS^\beta - PV^\beta + \sum_i \mu_i n_i^\beta. \tag{1.28}$$

The Helmholtz free energy F in the respective phase is obtained from the internal energy U by Legendre transformation:

$$F^\alpha = U^\alpha - TS^\alpha = -PV^\alpha + \sum_i \mu_i n_i^\alpha, \tag{1.29}$$

$$F^\beta = U^\beta - TS^\beta = -PV^\beta + \sum_i \mu_i n_i^\beta, \tag{1.30}$$

and

$$F^\sigma = U^\sigma - TS^\sigma = \gamma A + \sum_i \mu_i n_i^\sigma. \tag{1.31}$$

Furthermore, division of Eq. (1.31) by surface area A leads to

$$\gamma = f^\sigma - \sum_i \Gamma_i \mu_i, \tag{1.32}$$

where $f^\sigma = \frac{F^\sigma}{A}$ and $\Gamma_i = \frac{n_i^\sigma}{A}$ are the Helmholtz free energy per unit surface area and the adsorption of component i, respectively. In the one component system, γ is equal to f^σ by choosing the location of dividing surface at $\Gamma_1 = 0$. On the other hand, in the multicomponent systems, γ is not equal to f^σ since $\Gamma_{i\neq 1}$ is not zero, even if the location of dividing surface is chosen at $\Gamma_1 = 0$ [5]. Nevertheless, γ is independent of the location of the dividing surface, while f^σ is dependent of the location.

The Gibbs free energy G in the respective phase is obtained from the internal energy U, and the enthalpy $H(= U + PV)$ by two Legendre transformations:

$$G^\alpha = H^\alpha - TS^\alpha = \sum_i \mu_i n_i^\alpha, \tag{1.33}$$

$$G^{\beta} = H^{\beta} - TS^{\beta} = \sum_i \mu_i n_i^{\beta}, \tag{1.34}$$

and

$$G^{\sigma} = H^{\sigma} - TS^{\sigma} = \gamma A + \sum_i \mu_i n_i^{\sigma}. \tag{1.35}$$

It is reminded that G^{σ} is equal to F^{σ} since there is no surface volume in the Gibbs model. Consequently, Eq. (1.32) can be replaced by

$$\gamma = \frac{G^{\sigma}}{A} - \sum_i \Gamma_i \mu_i, \tag{1.36}$$

where $\frac{G^{\sigma}}{A}$ is the Gibbs free energy per unit surface area. The total Gibbs free energy G in the entire system is the sum of that in each phase:

$$\begin{aligned} G &= \gamma A + \sum_i \mu_i \left(n_i^{\alpha} + n_i^{\beta} + n_i^{\sigma} \right) \\ &= \gamma A + \sum_i \mu_i n_i. \end{aligned} \tag{1.37}$$

Equation (1.37) means that γ can be represented in terms of the Gibbs free energy in the entire system as follows:

$$\gamma = \left(\frac{\partial G}{\partial A} \right)_{T,P,n_1 \dots n_i}. \tag{1.38}$$

The Gibbs free energy is the most useful potential function under the experimental conditions at constant temperature and constant pressure. The complete differential of the Gibbs free energy of surface phase $G^{\sigma}\left(T, A, n_1^{\sigma}, \dots, n_i^{\sigma}\right)$ is

$$dG^{\sigma} = \left(\frac{\partial G^{\sigma}}{\partial T} \right)_{A,n_1^{\sigma} \dots n_i^{\sigma}} dT + \left(\frac{\partial G^{\sigma}}{\partial A} \right)_{T,n_1^{\sigma} \dots n_i^{\sigma}} dA + \sum_i \left(\frac{\partial G^{\sigma}}{\partial n_i^{\sigma}} \right)_{T,A,n_{j \neq i}^{\sigma}} dn_i^{\sigma}. \tag{1.39}$$

From the thermodynamic relationships of $\left(\frac{\partial G^{\sigma}}{\partial T}\right)_{A,n_1^{\sigma} \dots n_i^{\sigma}} = -S^{\sigma}$, $\left(\frac{\partial G^{\sigma}}{\partial A}\right)_{T,n_1^{\sigma} \dots n_i^{\sigma}} = \gamma$, and $\left(\frac{\partial G^{\sigma}}{\partial n_i^{\sigma}}\right)_{T,A,n_{j \neq i}^{\sigma}} = \mu_i$, Eq. (1.39) can be described as follows:

$$dG^{\sigma} = -S^{\sigma} dT + \gamma dA + \sum_i \mu_i dn_i^{\sigma}. \tag{1.40}$$

The complete differential of equation (1.35) is

$$dG^\sigma = \gamma dA + A d\gamma + \sum_i \mu_i dn_i^\sigma + \sum_i n_i^\sigma d\mu_i. \tag{1.41}$$

The following relation can be derived from the equality of Eqs. (1.40) and (1.41):

$$S^\sigma dT + A d\gamma + \sum_i n_i^\sigma d\mu_i = 0, \tag{1.42}$$

which is the Gibbs–Duhem equation for the surface plastically deformed. At constant temperature, Eq. (1.42) leads to

$$A d\gamma = -\sum_i n_i^\sigma d\mu_i. \tag{1.43}$$

Dividing both sides of Eq. (1.43) by A provides

$$d\gamma = -\sum_i \frac{n_i^\sigma}{A} d\mu_i = -\sum_i \Gamma_i d\mu_i, \tag{1.44}$$

which is named "the Gibbs adsorption isotherm". Equation (1.44) means that γ changes with μ_i. However, γ is independent of the location of dividing surface.

At constant temperature and pressure, the Gibbs–Duhem equations for the bulk α and β phases are

$$\sum_i x_i^\alpha d\mu_i = 0, \tag{1.45}$$

and

$$\sum_i x_i^\beta d\mu_i = 0, \tag{1.46}$$

where x_i^α and x_i^β are the mole fractions of component i in the α and β phases, respectively. By using Eqs. (1.45) and (1.46), $d\mu_1$ can be expressed as a function of $d\mu_{i\neq 1}$ and $x_{i\neq 1}^\alpha$ or $x_{i\neq 1}^\beta$:

$$d\mu_1 = -\sum_{i\neq 1} \frac{x_i^\alpha}{x_1^\alpha} d\mu_i = -\sum_{i\neq 1} \frac{x_i^\beta}{x_1^\beta} d\mu_i. \tag{1.47}$$

Moreover, Eq. (1.44) can be written as follows:

$$d\gamma = -\Gamma_1 d\mu_1 - \sum_{i\neq 1} \Gamma_i d\mu_i. \tag{1.48}$$

Substituting $d\mu_1$ of Eqs. (1.47) into (1.48) leads to

$$d\gamma = -\sum_{i \neq 1}\left(\Gamma_i - \Gamma_1 \frac{x_i^\alpha}{x_1^\alpha}\right) d\mu_i = -\sum_{i \neq 1}\left(\Gamma_i - \Gamma_1 \frac{x_i^\beta}{x_1^\beta}\right) d\mu_i. \quad (1.49)$$

The term of $\left(\Gamma_i - \Gamma_1 \frac{x_i^\alpha}{x_1^\alpha}\right)$ or $\left(\Gamma_i - \Gamma_1 \frac{x_i^\beta}{x_1^\beta}\right)$ in Eq. (1.49) corresponds to the relative surface excess $\Gamma_{i,1}$ of component i with respect to component 1.

Láng [3] indicated that the Gibbs adsorption isotherm can be expressed in terms of the relative surface excess of component i with respect to two selected components. If two components 1 and 2 are chosen, the following equations of $d\mu_1$ and $d\mu_2$ can be derived from Eqs. (1.45) to (1.46):

$$d\mu_1 = -\frac{x_2^\alpha}{x_1^\alpha} d\mu_2 - \sum_{i \neq 1,2} \frac{x_i^\alpha}{x_1^\alpha} d\mu_i, \quad (1.50)$$

and

$$d\mu_2 = -\frac{x_1^\beta}{x_2^\beta} d\mu_1 - \sum_{i \neq 1,2} \frac{x_i^\beta}{x_2^\beta} d\mu_i. \quad (1.51)$$

Equation (1.44) can be also written as follows:

$$d\gamma = -\Gamma_1 d\mu_1 - \Gamma_2 d\mu_2 - \sum_{i \neq 1,2} \Gamma_i d\mu_i. \quad (1.52)$$

Solving the set of Eqs. (1.50) and (1.51) for $d\mu_1$ and $d\mu_2$ and then substituting the results into Eq. (1.52), we obtain

$$d\gamma = -\sum_{i \neq 1,2}\left[\Gamma_i - \left(\frac{x_i^\alpha x_2^\beta - x_2^\alpha x_i^\beta}{x_1^\alpha x_2^\beta - x_2^\alpha x_1^\beta}\right)\Gamma_1 - \left(\frac{x_1^\alpha x_i^\beta - x_i^\alpha x_1^\beta}{x_1^\alpha x_2^\beta - x_2^\alpha x_1^\beta}\right)\Gamma_2\right] d\mu_i. \quad (1.53)$$

If the relative surface excess of component i with respect to two components 1 and 2 is defined as $\Gamma_{i,1,2}$,

$$\Gamma_{i,1,2} = \Gamma_i - \left(\frac{x_i^\alpha x_2^\beta - x_2^\alpha x_i^\beta}{x_1^\alpha x_2^\beta - x_2^\alpha x_1^\beta}\right)\Gamma_1 - \left(\frac{x_1^\alpha x_i^\beta - x_i^\alpha x_1^\beta}{x_1^\alpha x_2^\beta - x_2^\alpha x_1^\beta}\right)\Gamma_2, \quad (1.54)$$

and

$$d\gamma = -\sum_{i \neq 1,2} \Gamma_{i,1,2} d\mu_i, \quad (1.55)$$

are obtained [3]. Since the mole fractions of all components in the bulk α and β phases are fixed, $\Gamma_{i,1,2}$ is independent of the location of the dividing surface.

1.6 Surface Tension and Surface Stress

The reversible work dW_{p} required for creating new surface by dA under plastic deformation is defined by

$$dW_{\mathrm{p}} = \gamma \mathrm{dA}. \tag{1.56}$$

On the other hand, the reversible work dW_{e} for changing surface area by dA under elastic deformation is defined by

$$dW_{\mathrm{e}} = g\mathrm{dA}, \tag{1.57}$$

where g is named "surface stress." In Eq. (1.57), g as well as γ is a scalar quantity, which is valid only for an isotropic substance or for a crystal face with threefold, or greater, axis of symmetry. In the case of an anisotropic substance, surface stress is a tensor quantity g_{ij} as explained later.

Couchman et al. [6] derived the relationship between γ and g for the isotropic surface of one component system as follows. Provided that the total surface area A consists of the number N of surface atoms, the surface area a per surface atom is given by

$$a = \frac{A}{N}. \tag{1.58}$$

Since the location of the dividing surface can be chosen at $\Gamma_1 = 0$ in one component system, γ is equal to the Helmholtz free energy per unit surface area f^{σ} or the Gibbs free energy per unit surface area $\frac{G^{\sigma}}{A}$ (see Eq. (1.32) or Eq. (1.36)). If the excess free energy per surface atom is denoted by φ, γ is represented by

$$\gamma = \frac{\varphi N}{A} = \frac{\varphi}{a}. \tag{1.59}$$

When the initial surface with A and N is changed by $A + dA$ and $N + dN$ at constant temperature, the net change of surface area dA consists of the change of number of surface atoms dN due to partly plastic deformation and of the change of area per surface atom da due to partly elastic deformation. The corresponding work δW is given by

$$\delta W = g' dA = \delta(\varphi N), \tag{1.60}$$

where g' is named "effective" surface stress by Couchman et al. [6], and it takes an intermediate value of γ and g. Furthermore, Eq. (1.60) can be transformed to

$$\delta W = \varphi \frac{\partial N}{\partial A} dA + N \frac{\partial \varphi}{\partial A} dA. \tag{1.61}$$

From the comparison between Eqs. (1.60) and (1.61), g' is given by

$$g' = \varphi \frac{\partial N}{\partial A} + N \frac{\partial \varphi}{\partial A}. \tag{1.62}$$

In the case where the surface is subjected to plastic deformation like liquid, the following relationships hold:

$$\left(\frac{\partial \varphi}{\partial A} \right)_a = 0, \tag{1.63}$$

and

$$\left(\frac{\partial N}{\partial A} \right)_a = \frac{1}{a}. \tag{1.64}$$

Substituting Eqs. (1.63) and (1.64) into Eq. (1.62), we obtain

$$g' = \varphi \left(\frac{\partial N}{\partial A} \right)_a = \frac{\varphi}{a} = \gamma, \tag{1.65}$$

which is identical with Eq. (1.59). In the case where the surface is subjected to elastic deformation like solid, the following relationships also hold:

$$\frac{\partial N}{\partial A} = 0, \tag{1.66}$$

and

$$dA = N da. \tag{1.67}$$

Therefore, Eq. (1.62) is transformed to

$$g' = N \left(\frac{\partial \varphi}{\partial A} \right)_N = \left(\frac{\partial \varphi}{\partial a} \right)_N = g. \tag{1.68}$$

The substitution of $\varphi = \gamma a$ (see Eq. (1.59)) into Eq. (1.68) leads to

$$g = \gamma + \left(a \frac{\partial \gamma}{\partial a} \right)_N = \gamma + \left(\frac{\partial \gamma}{\partial \varepsilon} \right)_N. \tag{1.69}$$

In Eq. (1.69), $\partial\varepsilon = \frac{\partial a}{a}$ is the elastic strain and corresponds to the second term in the right-hand side of Eq. (1.22). Equation (1.69) is named "Shuttleworth equation" since Shuttleworth [7] derived first this equation for an isotropic solid surface.

In the case where the surface is subjected to partly plastic and elastic deformation, the substitution of $\varphi = \gamma a$ into Eq. (1.62) leads to

$$\begin{aligned} g' &= \gamma a \frac{\partial N}{\partial A} + N \frac{\partial(\gamma a)}{\partial A} \\ &= \gamma \frac{\partial(aN)}{\partial A} + aN \frac{\partial \gamma}{\partial A} \\ &= \gamma + A \frac{\partial \gamma}{\partial A} = \gamma + \frac{\partial \gamma}{\partial \varepsilon'}, \end{aligned} \quad (1.70)$$

where $\partial\varepsilon'$ is the sum of plastic strain $\left(\frac{\partial N}{N}\right)$ and elastic strain $\left(\frac{\partial a}{a}\right)$ as shown in Eq. (1.22). For an anisotropic solid surface, surface stress is a tensor quantity and the tensor equivalent of the Shuttleworth equation [8, 9] is represented by

$$g_{nm} = \gamma \delta_{nm} + \frac{\partial \gamma}{\partial \varepsilon_{nm}}, \quad (1.71)$$

where δ_{nm} is the Kronecker delta and ε_{nm} is the strain tensor of the elastic deformation.

The Kronecker delta has the following property:

$$\delta_{nm} = 1 \text{ at } n = m, \quad (1.72)$$

and

$$\delta_{nm} = 0 \text{ at } n \neq m. \quad (1.73)$$

The surface stress g_{nm} (J m^{-2}), i.e., the reversible work required to form unit area of new surface under elastic deformation, corresponds to a force per unit length (N m^{-1}), acting in the mth direction on an edge normal to the nth direction (n and m being in the plane of the surface) at constant temperature and chemical potential. For an isotropic substance, g_{nn} is equal to g_{mm}, and g_{nm} (or g_{mn}) reduces to zero in Eq. (1.71), such that the surface stress becomes the scalar quantity denoted by g. Figure 1.3 shows the components of surface stress tensor (g_{xx}, g_{yy}, g_{xy}, and g_{yx}) acting on each edge of the x–y surface plane [9]. The derivation of Eq. (1.71) has been made by Mullins [10] and Linford [11]. We explain the detailed derivation of Eq. (1.71) by Linford [11] in Appendix 1 of this chapter.

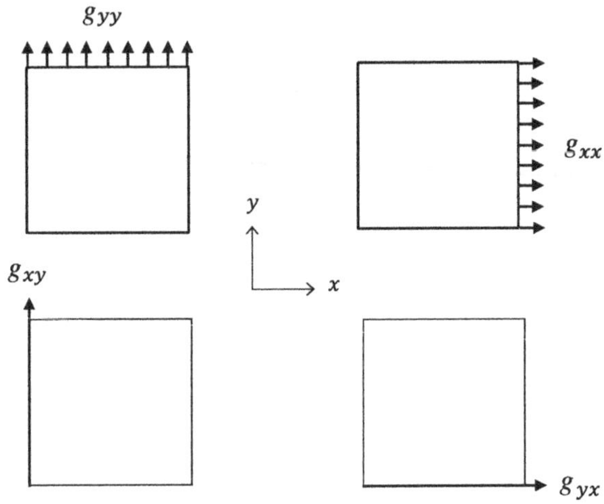

Fig. 1.3 Components of surface stress tensor (g_{xx}, g_{yy}, g_{xy}, and g_{yx}) acting on each edge of the x–y surface plane [9]. Reprinted with the permission from [9], Copyright 1978, American Chemical Society

1.7 Gibbs–Duhem Equation of Solid Surface

According to Couchman et al. [6, 12–14], for an isotropic solid surface subjected to partly plastic and elastic deformation, the differential of the excess internal energy of the surface phase may be written in place of Eq. (1.25) as follows:

$$dU^{\sigma} = TdS^{\sigma} + g'dA + \sum_i \mu_i dn_i^{\sigma}. \quad (1.74)$$

Under a constant elastic strain, i.e., $d\varepsilon = 0$, the reversible work required for stretching the surface area by A would be γA because of γ being independent of the area under the plastic deformation. The excess internal energy of the surface phase U^{σ} is a state function which is determined only by the initial and final states and does not depend any routes from the initial to the final state (e.g., the route under a constant elastic strain), and thus U^{σ} may be represented by the same equation as Eq. (1.26). The differential of Eq. (1.26) is

$$dU^{\sigma} = TdS^{\sigma} + S^{\sigma}dT + \gamma dA + Ad\gamma + \sum_i \mu_i dn_i^{\sigma} + \sum_i n_i^{\sigma} d\mu_i. \quad (1.75)$$

Subtracting Eqs. (1.74) from (1.75), we obtain

$$\frac{S^{\sigma}}{A}dT + d\gamma + (\gamma - g')d\varepsilon' + \sum_i \Gamma_i d\mu_i = 0, \quad (1.76)$$

where $d\varepsilon' = \frac{dA}{A}$ and $\Gamma_i = \frac{n_i^\sigma}{A}$. If the effective surface stress g' is conjugated to the general change (partly plastic and elastic) in surface area, g' may be formally defined in terms of γ and g by the following equation [14]:

$$g' = \frac{d\varepsilon_p}{d\varepsilon'}\gamma + \frac{d\varepsilon}{d\varepsilon'}g, \tag{1.77}$$

where $d\varepsilon_\mathrm{p} = \frac{dN}{N}$ and $d\varepsilon = \frac{da}{a}$ are the plastic and elastic contributions to the total strain $d\varepsilon' = \frac{dA}{A}$ as shown in Eq. (1.22). The substitution of Eqs. (1.22) and (1.77) into Eq. (1.76) gives

$$\frac{S^\sigma}{A}dT + d\gamma + (\gamma - g)d\varepsilon + \sum_i \Gamma_i d\mu_i = 0. \tag{1.78}$$

In the case of an anisotropic solid surface, Eq. (1.78) is transformed to

$$\frac{S^\sigma}{A}dT + d\gamma + (\gamma\delta_{nm} - g_{nm})d\varepsilon_{nm} + \sum_i \Gamma_i d\mu_i = 0. \tag{1.79}$$

Equation (1.78) or Eq. (1.79) may be regarded as the general form of the Gibbs–Duhem equation of solid surfaces.

At constant temperature and chemical potential, the expression similar to the Shuttleworth equation in Eq. (1.69) or Eq. (1.71) is obtained:

$$g = \gamma + \left(\frac{\partial\gamma}{\partial\varepsilon}\right)_{T,\mu_i}, \tag{1.80}$$

or

$$g_{nm} = \gamma\delta_{nm} + \left(\frac{\partial\gamma}{\partial\varepsilon_{nm}}\right)_{T,\mu_i}. \tag{1.81}$$

At constant temperature, the Gibbs adsorption isotherm is derived from Eq. (1.78) or Eq. (1.79):

$$d\gamma = -\sum_i \Gamma_i d\mu_i + (g - \gamma)d\varepsilon, \tag{1.82}$$

or

$$d\gamma = -\sum_i \Gamma_i d\mu_i + (\gamma\delta_{nm} - g_{nm})d\varepsilon_{nm}. \tag{1.83}$$

Furthermore, the surface excess of component i at constant elastic strain ($d\varepsilon = 0$) is expressed by

$$\Gamma_i = -\left(\frac{\partial \gamma}{\partial \mu_i}\right)_{T,\varepsilon,\mu_{j\neq i}}, \tag{1.84}$$

or

$$\Gamma_i = -\left(\frac{\partial \gamma}{\partial \mu_i}\right)_{T,\varepsilon_{nm},\mu_{j\neq i}}. \tag{1.85}$$

1.8 Electrified Interface and Electrocapillarity

When a solid such as metal is immersed in an electrolyte solution, the solid/solution interface, i.e., solid electrode surface, is electrified. The derivation of the thermodynamic parameters for the electrified interface is not simple because the electrochemical potential $\tilde{\mu}_i$ in place of the chemical potential μ_i has to be used for the charged particles such as electrons and electrolyte ions. The thermodynamic parameters for the electrified interface have been rigorously derived by Parsons [15] and recently by Láng [3]. The total differential of the excess internal energy for the electrified interface of a liquid metal such as mercury is given by

$$dU^\sigma = TdS^\sigma + \gamma dA + \sum_i \tilde{\mu}_i^\sigma dn_i^\sigma. \tag{1.86}$$

The electrochemical potential of component i is the sum of the chemical term (μ_i) and electrostatic term ($z_i F \Phi$):

$$\tilde{\mu}_i = \mu_i + z_i F \Phi, \tag{1.87}$$

where Φ is the inner potential, z_i, the charge number with plus or minus sign of particles, and F is the Faraday constant. The Gibbs–Duhem equation can be written as

$$S^\sigma dT + Ad\gamma + \sum_i n_i^\sigma d\tilde{\mu}_i^\sigma = 0. \tag{1.88}$$

At constant temperature, the Gibbs adsorption isotherm of the electrified interface is obtained:

$$d\gamma = -\sum_i \frac{n_i^\sigma}{A} d\tilde{\mu}_i = -\sum_i \Gamma_i d\tilde{\mu}_i. \tag{1.89}$$

Let us consider the simple case where mercury (α phase) is in contact with aqueous solution (β phase) containing a single salt of KCl. In the β phase, KCl dissociates into cations K^+ and anions Cl^- in undissociated water solvent. The electrode is regarded as an ideally polarizable electrode. This means that no charge transfer reaction such as a redox reaction takes place across the interface and the interface is only charged to form an electric double layer. Suppose also that the potential of the mercury electrode is controlled with respect to an Ag/AgCl reference electrode. The electrochemical cell may be described as follows [16]:

$$\mathrm{Cu} \mid \mathrm{Hg} \mid \mathrm{K}^+, \mathrm{Cl}^-, \mathrm{H_2O} \mid \mathrm{Ag} \mid \mathrm{AgCl} \mid \mathrm{Cu}', \tag{1.90}$$

where Cu and Cu′ are copper wires electrically contacted with the Hg and Ag/AgCl reference electrodes, respectively. A surface excess of electrons on the mercury can be regarded as excess of charge on the electrode surface [16].

The Gibbs adsorption isotherm can be expressed by separating two terms of the components of the mercury electrode and of the solution:

$$d\gamma = -\left(\Gamma_{\mathrm{Hg}} d\tilde{\mu}_{\mathrm{Hg}} + \Gamma_e d\tilde{\mu}_{e(\mathrm{Hg})}\right) - \left(\Gamma_{\mathrm{K}^+} d\tilde{\mu}_{\mathrm{K}^+} + \Gamma_{\mathrm{Cl}^-} d\tilde{\mu}_{\mathrm{Cl}^-} + \Gamma_{\mathrm{w}} d\tilde{\mu}_{\mathrm{w}}\right), \tag{1.91}$$

where $\tilde{\mu}_{e(\mathrm{Hg})}$ refers to electrons in the mercury phase and the subscript w means water as undissociated solvent. There are the following relationships between electrochemical potentials:

$$\Gamma_e d\tilde{\mu}_{e(\mathrm{Hg})} = \Gamma_e d\tilde{\mu}_{e(\mathrm{Cu})}, \tag{1.92}$$

$$\tilde{\mu}_{\mathrm{KCl}} = \tilde{\mu}_{\mathrm{K}^+} + \tilde{\mu}_{\mathrm{Cl}^-} = \mu_{\mathrm{KCl}}, \tag{1.93}$$

and

$$\tilde{\mu}_{\mathrm{w}} = \mu_{\mathrm{w}}. \tag{1.94}$$

Since $d\tilde{\mu}_{\mathrm{Hg}} = d\mu_{\mathrm{Hg}} = 0$, Eq. (1.91) can be transformed to

$$d\gamma = -\Gamma_e d\tilde{\mu}_{e(\mathrm{Cu})} - \left(\Gamma_{\mathrm{K}^+} d\mu_{\mathrm{KCl}} - \Gamma_{\mathrm{K}^+} d\tilde{\mu}_{\mathrm{Cl}^-} + \Gamma_{\mathrm{Cl}^-} d\tilde{\mu}_{\mathrm{Cl}^-} + \Gamma_{\mathrm{w}} d\mu_{\mathrm{w}}\right). \tag{1.95}$$

The Ag/AgCl electrode is reversible with respect to Cl^- ions in the aqueous β phase so that the following equilibrium holds at the reference interface:

$$\tilde{\mu}_{\mathrm{AgCl}} + \tilde{\mu}_{e(\mathrm{Cu}')} = \tilde{\mu}_{\mathrm{Ag}} + \tilde{\mu}_{\mathrm{Cl}^-}. \tag{1.96}$$

Furthermore, from $d\tilde{\mu}_{\mathrm{AgCl}} = d\tilde{\mu}_{\mathrm{Ag}} = 0$, the differential of Eq. (1.96) is given by

$$d\tilde{\mu}_{e(\mathrm{Cu}')} = d\tilde{\mu}_{\mathrm{Cl}^-}. \tag{1.97}$$

The substitution of Eqs. (1.97) into (1.95) leads to

$$d\gamma = -\Gamma_e d\tilde{\mu}_{e(\mathrm{Cu})} + (\Gamma_{\mathrm{K}^+} - \Gamma_{\mathrm{Cl}^-})d\tilde{\mu}_{e(\mathrm{Cu}')} - (\Gamma_{\mathrm{K}^+} d\mu_{\mathrm{KCl}} + \Gamma_{\mathrm{w}} d\mu_{\mathrm{w}}). \quad (1.98)$$

The surface charge density q on the metal side of the interface is

$$q = -F\Gamma_e. \quad (1.99)$$

From the condition of electroneutrality for the interface, the charge density on the solution side is equal to $-q$, that is,

$$-q = F(\Gamma_{\mathrm{K}^+} - \Gamma_{\mathrm{Cl}^-}). \quad (1.100)$$

Moreover,

$$d\tilde{\mu}_{e(\mathrm{Cu})} - d\tilde{\mu}_{e(\mathrm{Cu}')} = -Fd\left(\Phi^{\mathrm{Cu}} - \Phi^{\mathrm{Cu}'}\right) = -FdE_-, \quad (1.101)$$

where Φ^{Cu} and $\Phi^{\mathrm{Cu}'}$ are the inner potentials of Cu and Cu′, respectively, and E_- is the potential of the mercury electrode with respect to the reference electrode. From Eqs. (1.99), (1.100), and (1.101), Eq. (1.98) can be converted to

$$d\gamma = -(\Gamma_{\mathrm{K}^+} d\mu_{\mathrm{KCl}} + \Gamma_{\mathrm{w}} d\mu_{\mathrm{w}}) - qdE_-. \quad (1.102)$$

The Gibbs–Duhem relation for the aqueous β phase at constant temperature and pressure is

$$x^{\beta}_{\mathrm{KCl}} d\mu_{\mathrm{KCl}} + x^{\beta}_{\mathrm{w}} d\mu_{\mathrm{w}} = 0, \quad (1.103)$$

where x^{β}_{KCl} and x^{β}_{w} are the mole fractions of KCl and water in the β phase, respectively. The elimination of $d\mu_{\mathrm{w}}$ by substituting Eqs. (1.103) into (1.102) gives

$$d\gamma = -\left(\Gamma_{\mathrm{K}^+} - \frac{x^{\beta}_{\mathrm{KCl}}}{x^{\beta}_{\mathrm{w}}}\Gamma_{\mathrm{w}}\right) d\mu_{\mathrm{KCl}} - qdE_-, \quad (1.104)$$

where $\left(\Gamma_{\mathrm{K}^+} - \frac{x^{\beta}_{\mathrm{KCl}}}{x^{\beta}_{\mathrm{w}}}\Gamma_{\mathrm{w}}\right)$ is the relative surface excess of K^+ ions and is independent of the location of the dividing surface. If the relative surface excess of K^+ ions is represented as

$$\Gamma_{\mathrm{K}^+,\mathrm{w}} = \Gamma_{\mathrm{K}^+} - \frac{x^{\beta}_{\mathrm{KCl}}}{x^{\beta}_{\mathrm{w}}}\Gamma_{\mathrm{w}}, \quad (1.105)$$

the Gibbs adsorption isotherm can be eventually written as follows:

$$d\gamma = -\Gamma_{\mathrm{K}^+,\mathrm{w}} d\mu_{\mathrm{KCl}} - qdE_-. \quad (1.106)$$

In the case where the reference electrode which is reversible with respect to K^+ ions, is employed, the Gibbs adsorption isotherm can be represented by

$$d\gamma = -\left(\Gamma_{Cl^-} - \frac{x_{KCl}^{\beta}}{x_W^{\beta}}\Gamma_W\right)d\mu_{KCl} - qdE_+, \tag{1.107}$$

or

$$d\gamma = -\Gamma_{Cl^-,W}d\mu_{KCl} - qdE_+. \tag{1.108}$$

Equation (1.106) or Eq. (1.108) is named "electrocapillary equation." At constant T and μ_{KCl}, the Lippmann equation can be obtained from Eq. (1.106) or Eq. (1.108):

$$\left(\frac{\partial\gamma}{\partial E_{\pm}}\right)_{T,\mu_{KCl}} = -q. \tag{1.109}$$

The general form of the Gibbs–Duhem equation for a solid electrode (electrified interface) [9, 17, 18] may be written as

$$\frac{S^{\sigma}}{A}dT + d\gamma + qdE + (\gamma\delta_{nm} - g_{nm})d\varepsilon_{nm} + \sum_i \Gamma_i d\mu_i = 0. \tag{1.110}$$

Equation (1.110) is equivalent to the case where the electrostatic term of qdE is added to the left-hand side of Eq. (1.79). In Eq. (1.110), qdE (in place of qdE_+ or qdE_-) is employed irrespective of the reference electrode, and thus, the content of $\sum_i \Gamma_i d\mu_i$ may be modified as shown in Eq. (1.106) or Eq. (1.108), depending on the kinds of the reference electrode employed in experiment. At constant temperature, the electrocapillary equation is obtained from Eq. (1.110):

$$d\gamma = -\sum_i \Gamma_i d\mu_i - qdE + (g_{nm} - \gamma\delta_{nm})d\varepsilon_{nm}. \tag{1.111}$$

In the case of a liquid electrode such as mercury, the term of $(g_{nm} - \gamma\delta_{nm})d\varepsilon_{nm}$ vanishes since g is equal to γ, and thereby, Eq. (1.111) is identical with Eq. (1.106) or Eq. (1.108). The generalized Lippmann equation can be obtained from Eq. (1.111):

$$\left(\frac{\partial\gamma}{\partial E}\right)_{T,\mu_i} = -q + (g_{nm} - \gamma\delta_{nm})\left(\frac{\partial\varepsilon_{nm}}{\partial E}\right)_{T,\mu_i}. \tag{1.112}$$

For a liquid electrode, the last term in the right-hand side of Eq. (1.112) vanishes. However, some discussions are needed upon the magnitude of the last term for a solid electrode.

Let estimate the magnitude of the last term for an unreconstructed Au (111) electrode (see Sect. 3.2 of Chap. 3 for reconstructed or unreconstructed surfaces) which is typical of an isotropic solid. Needs and Mansfield [19], and Payne et al. [20] calculated the values of $g = 2.77\ \mathrm{J\ m^{-2}}$ and $\gamma = 1.25\ \mathrm{J\ m^{-2}}$ for the unreconstructed Au (111) surface by using pseudopotential total energy method. Consequently, the magnitude of $(g - \gamma)$ is estimated to be about 1.5 J $\mathrm{m^{-2}}$. Lipkowski et al. [21] calculated the magnitude of surface elastic strain from the curvature change of a cantilever bending used for the surface stress measurement of the unreconstructed Au (111) electrode in $HClO_4$ solution [22]. The procedure of the calculation is described in Appendix 2 of this chapter. As a result, the magnitude of $\left(\frac{\partial \varepsilon}{\partial E}\right)_{T,P,\mu_i} \approx 5 \times 10^{-8}$ $\mathrm{V^{-1}}$ is calculated for the unreconstructed Au (111) electrode [21]. The magnitude of the last term in the right-hand side of Eq. (1.112) is eventually estimated to be [21]:

$$(g - \gamma)\left(\frac{\partial \varepsilon}{\partial E}\right)_{T,\mu_i} \approx 7.5 \times 10^{-8} \mathrm{C\ m^{-2}}. \tag{1.113}$$

The absolute value of q is around 0.4 C $\mathrm{m^{-2}}$ for the Au (111) electrode in the potential region of electric double layer [23, 24] where no significant charge transfer occurs across the interface. Therefore, the last term in the right-hand side of Eq. (1.112) is negligibly small compared to the first term. This means that the Lippmann equation expressed by Eq. (1.109) for a liquid electrode is also valid for a solid electrode.

At constant elastic strain, the Lippmann equation for an isotropic solid electrode is obtained from Eq. (1.111):

$$\left(\frac{\partial \gamma}{\partial E}\right)_{T,\mu_i,\varepsilon} = -q. \tag{1.114}$$

At constant T, μ_i, and E for an isotropic solid electrode, Eq. (1.111) leads to the following relationship:

$$\left(\frac{\partial \gamma}{\partial \varepsilon}\right)_{T,\mu_i,E} = g - \gamma. \tag{1.115}$$

Since Eq. (1.111) is an exact differential under the restrictions of an ideally polarizable, elastically strained electrode [25, 26], $d\gamma$ at constant T and μ_i, can be written by

$$d\gamma = \left(\frac{\partial \gamma}{\partial E}\right)_{T,\mu_i,\varepsilon} dE + \left(\frac{\partial \gamma}{\partial \varepsilon}\right)_{T,\mu_i,E} d\varepsilon. \tag{1.116}$$

The Maxwell relation holds between the second mixed partial derivatives of Eq. (1.116):

$$\frac{\partial}{\partial \varepsilon}\left[\left(\frac{\partial \gamma}{\partial E}\right)_{T,\mu_i,\varepsilon}\right]_E = \frac{\partial}{\partial E}\left[\left(\frac{\partial \gamma}{\partial \varepsilon}\right)_{T,\mu_i,E}\right]_\varepsilon. \tag{1.117}$$

The substitution of Eqs. (1.114) and (1.115) into the corresponding square brackets in Eq. (1.117) leads to the following equations:

$$-\left(\frac{\partial q}{\partial \varepsilon}\right)_E = \left(\frac{\partial g}{\partial E}\right)_\varepsilon - \left(\frac{\partial \gamma}{\partial E}\right)_\varepsilon = \left(\frac{\partial g}{\partial E}\right)_\varepsilon + q, \tag{1.118}$$

or

$$\left(\frac{\partial g}{\partial E}\right)_\varepsilon = -q - \left(\frac{\partial q}{\partial \varepsilon}\right)_E. \tag{1.119}$$

Equation (1.119) was first derived by Gokhshtein [27] and so called "Gokhshtein equation." The validity of the Gokhshtein equation has been confirmed by Valincius [25] and Proost [26].

The differential of the internal energy for the electrified interface subjected to elastic deformation can be written by

$$dU^\sigma = TdS^\sigma + EdQ + gdA_\mathrm{e} + \sum_i \mu_i dn_i^\sigma, \tag{1.120}$$

where dA_e is the change of surface area due to elastic deformation. Valincius [25] introduced the function Ψ^σ as follows:

$$\Psi^\sigma = U^\sigma - TS^\sigma - \sum_i \mu_i n_i^\sigma. \tag{1.121}$$

The function Ψ^σ as well as U^σ is a homogeneous function of the first order with respect to all variable. The differential of the function Ψ^σ at constant T and μ_i is

$$\mathrm{d}\Psi^\sigma = EdQ + gdA_\mathrm{e}. \tag{1.122}$$

Euler's criteria for Eq. (1.122) is

$$\left(\frac{\partial g}{\partial Q}\right)_{A_\mathrm{e}} = \left(\frac{\partial E}{\partial A_\mathrm{e}}\right)_Q. \tag{1.123}$$

Furthermore, Eq. (1.123) can be expressed in terms of surface charge density q and elastic strain ε:

$$\left(\frac{\partial g}{\partial q}\right)_\varepsilon = \left(\frac{\partial E}{\partial \varepsilon}\right)_q. \tag{1.124}$$

Equation (1.124) was also first derived by Gokhshtein [27] who developed a piezo-electric technique for measuring separately the derivatives in the left- and right-hand sides of Eq. (1.124) and confirmed the validity of Eq. (1.124). The piezoelectric technique for detecting the derivative of surface stress change is explained in Sect. 2.2 of Chap. 2. The electro-chemo-mechanical properties of solid electrode surfaces are directly associated with the thermodynamic equations of the solid electrode surfaces. It is emphasized that the potential- or charge-induced surface stress change and the surface elastic strain-induced potential or charge change are linked with Eqs. (1.119) and (1.124).

1.9 Electrocapillary Curves of Liquid and Solid Metal Electrodes

In the simplest model of an electrified metal/solution interface, a layer of solvation ions on the outer Helmholtz plane (OHP) constitutes the entire excess charge in the solution side which has a sign opposite to that on the metal side; i.e., two layers of excess charge (electric double layer) behave like a parallel-plate condenser. The OHP is defined as the location of centers of solvation ions which can approach most closely to the metal side. Since the electric double-layer capacity c_o in the simplest model is independent of potential, the surface charge density q on the metal side is given by

$$q = c_o(E - E_{pzc}), \tag{1.125}$$

where E_{pzc} is the potential at which q becomes zero, i.e., the potential of zero charge.

For a liquid metal electrode such as mercury, the changes in surface tension can be obtained by the integration of the Lippmann equation of Eq. (1.109) with respect to potential:

$$\gamma - \gamma_{pzc} = -\frac{1}{2}c_o(E - E_{pzc})^2 = -\frac{q^2}{2c_o}, \tag{1.126}$$

where γ_{pzc} is the surface tension at E_{pzc} and γ takes a maximum at $q = 0$, i.e., at E_{pzc}.

The solid curve in Fig. 1.4 illustrates the electrocapillary curve (γ vs. E curve) calculated from Eq. (1.126) by using $c_o = 0.20\ \mathrm{F\ m^{-2}}$ and $\gamma_{pzc} = 0.426\ \mathrm{J\ m^{-2}}$. The electrocapillary curve is a perfect parabola symmetrical at E_{pzc}. The value of $\gamma_{pzc} = 0.426\ \mathrm{J\ m^{-2}}$ corresponds to that of γ_{pzc} for the mercury electrode in 0.01 M KF solution [28]. The maximum of the electrocapillary curve at E_{pzc} is named "electrocapillary maximum (ecm)." However, the electrocapillary curve obtained experimentally deviates inwards from a parabolic shape, particularly in the potential region more positive than E_{pzc} as shown in the dotted curve of Fig. 1.4. The dotted curve was calculated by changing c_o from 0.20 to 0.30 F m^{-2} at $E > E_{pzc}$. The increase in c_o enhances the inward deviation of the electrocapillary curve. The deviation of the

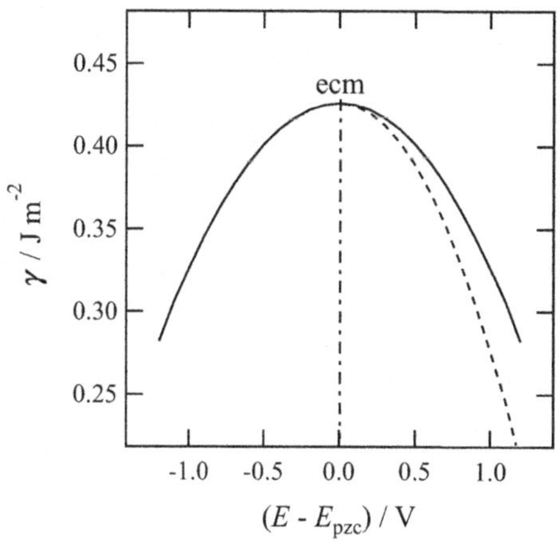

Fig. 1.4 Electrocapillary curve (γ vs. E curve) of a liquid metal electrode such as mercury. The solid curve was calculated from Eq. (1.128) by using $c_o = 0.20$ F m^{-2} and $\gamma_{pzc} = 0.426$ J m^{-2}, while the dotted curve was calculated by changing c_o from 0.20 to 0.30 F m^{-2} at $E > E_{pzc}$. The value of $\gamma_{pzc} = 0.426$ J m^{-2} corresponds to that of γ_{pzc} for the mercury electrode in 0.01 M KF solution [28]

measured electrocapillary curve from the parabolic shape means that the capacity of the electric double layer depends on potential. For the potential dependence of the capacity, it is convenient to use the differential capacity c defined by

$$c = \left(\frac{\partial q}{\partial E}\right)_{T,\mu_i}. \tag{1.127}$$

The combination of Eq. (1.127) with Eq. (1.109) leads to

$$c = -\left(\frac{\partial^2 \gamma}{\partial E^2}\right)_{T,\mu_i}. \tag{1.128}$$

If the differential capacity is measured as a function of potential, the values of q and γ can be obtained by integration:

$$q = \int_{E_{pzc}}^{E} c\, dE, \tag{1.129}$$

and

$$\gamma = \gamma_{pzc} - \int_{E_{pzc}}^{E} q\, dE, \tag{1.130}$$

or

$$\gamma = \gamma_{\text{pzc}} - \int_{E_{\text{pzc}}}^{E} \int c \, dE. \tag{1.131}$$

The potential dependence of the electric double-layer capacity indicates that the structure of the double layer is not simple like a parallel-plate condenser. The metal electrode surface is largely covered with adsorbed water molecules. Since a water molecule has a dipole, the orientation of adsorbed water dipoles depends on the sign and magnitude of the surface charge on the metal side which are controlled by the electrode potential. Furthermore, the anions such as Cl^- and Br^- as compared to the cations such as Na^+ and K^+ strip easily their solvation (hydration) sheaths, repel adsorbed water molecules away from electrode sites, and come into contact with a bare electrode. The location of centers of these adsorbed ions is defined as the inner Helmholtz plane (IHP). The adsorption of dehydrated ions at the IHP is termed "contact (or specific) adsorption." The hydration is much dependent on the radii of the ions. The smaller ions (Na^+, K^+, and F^-) are tightly wrapped up in the solvation sheaths. Therefore, the contact or specific adsorption is not expected for the smaller ions. In addition, the behavior of ions far from the electrode surface is affected by thermal and electric forces to form a diffuse-charge layer with ionic distribution in the outside of the OHP. The thickness of the diffuse-charge layer increases with decreasing electrolyte concentration. The electric double-layer capacity is influenced to some extent by the diffuse-charge layer when the electrolyte concentration is low. In contrast to the diffuse-charge layer, the layer of IHP and OHP is named "compact layer."

As shown in Fig. 1.5, the surface tension of the mercury electrode can be measured by a capillary electrometer [29] in which the force due to the weight of the mercury column in a fine glass tube is exactly balanced with the total force due to the surface tension at the perimeter of the contact of mercury with the inner wall of the glass capillary and electrolyte solution:

$$2\pi r \Gamma \cos\theta = \pi r^2 h \rho g, \tag{1.132}$$

where r is the radius of the inner wall of glass capillary, θ is the contact angle, h is the height of the mercury column, ρ is the density of mercury, and g is the acceleration of gravity. The term in the right-hand side of Eq. (1.132) is the force due to the weight of the mercury column, while the term in the left-hand side of Eq. (1.132) is the total force due to the surface tension. In the case of $\theta \approx 0$ at the mercury/glass/solution interface, the surface tension is obtained by

$$\gamma = \frac{rh\rho g}{2}. \tag{1.133}$$

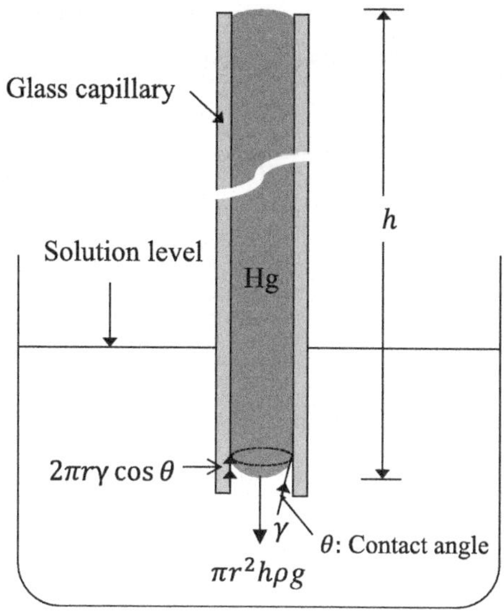

Fig. 1.5 Schema of a capillary electrometer for measurement of the surface tension of the mercury electrode [29]. The symbols shown are r: the radius of the inner wall of glass capillary; θ: the contact angle; h: the height of the mercury column; ρ: the density of mercury; and g: the acceleration of gravity. Reprinted from [29], Copyright 1970, Plenum Press, New York, with permission from Springer Nature

Here, it is reminded that the symbol of g in Eqs. (1.132) and (1.133) is not assigned to the surface stress.

Figure 1.6 shows the electrocapillary curves measured for the mercury electrode in electrolyte solutions containing various anions [30]. The potential in the abscissa of Fig. 1.6 is referred to E_{pzc} of the mercury electrode in NaF solution in which no contact adsorption occurs. The electrocapillary curves merge in the potential region more negative than $E - E_{pzc} = -0.70$ V, irrespective of anion species. On the other hand, in the potential region more positive than $E - E_{pzc} = -0.50$ V, the inward deviation of the electrocapillary curves from a parabolic shape increases significantly in the order of $OH^- < Cl^- < Br^- < I^-$, in response to the magnitude of the free energy change for contact adsorption of the anions on mercury. The negative shift of E_{pzc} due to the contact adsorption of the anions, accompanying the reduction of ecm, means that the strong chemical affinity of anions with mercury induces the contact adsorption even if the mercury electrode side is negatively charged. Figure 1.7 shows the solvent adsorption model at the negatively charged mercury electrode [29], which is useful for understanding the contact adsorption of anions. In Fig. 1.7, the diffuse-charge layer is added to the model proposed by Bockris et al. [31].

At present, it is unable to measure directly the electrocapillary curve of a solid metal electrode since no tools for measuring the surface tension of the solid electrode as a function of potential have been developed so far. Nevertheless, if the surface charge density or the differential capacity in addition to the position of E_{pzc} is known

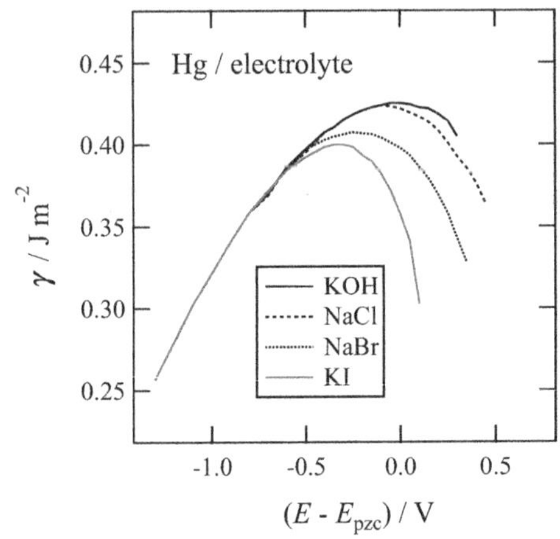

Fig. 1.6 Electrocapillary curves measured for the mercury electrode in electrolyte solutions containing various anions [30]. Reprinted with the permission from [30], Copyright 1947, American Chemical Society

as a function of potential, the electrocapillary curve of the solid electrode can be obtained by using Eq. (1.130) or Eq. (1.131) under the condition that the last term in the right-hand side of Eq. (1.112); i.e., $(g - \gamma)\left(\frac{\partial \varepsilon}{\partial E}\right)_{T,\mu_i}$ is neglected. Figure 1.8 shows the surface charge density versus potential (q vs. E) curves for an unreconstructed Au (111) electrode in 0.1 M $HClO_4$ solutions without and with 10^{-5} M, and 5×10^{-3} M K_2SO_4 [32]. The potential in the abscissa of Fig. 1.8 is referred to a saturated calomel electrode (SCE), the potential of which is more positive by 0.241 V than that of a standard hydrogen electrode (SHE). Since the absolute value of γ for a solid electrode cannot be measured experimentally, the integration of the charge density data (in Fig. 1.8) with respect to potential is made from −0.20 V (SCE) to the positive direction to obtain the changes in surface tension $\Delta\gamma$, referred to zero at −0.20 V (SCE). Figure 1.9 shows the electrocapillary curves ($\Delta\gamma$ vs. E) thus obtained for the unreconstructed Au (111) electrode [32]. The electrocapillary curves merge at $E < 0.20$ V (SCE) and diverge at more positive potentials where γ decreases with increasing bulk SO_4^{2-} concentration, indicating that the adsorption of sulfate proceeds significantly at $E > 0.20$ V (SCE). As explained above, the electrocapillary curve ($\Delta\gamma$ vs. E) for a solid electrode can be obtained from the data of the differential capacity or surface charge density measured as a function of potential.

1.10 Surface Stress versus Potential Curve of Solid Metal Electrode

It is unable to measure the absolute value of the surface stress as well as the surface tension for a solid metal electrode. Nevertheless, the changes in surface stress can be

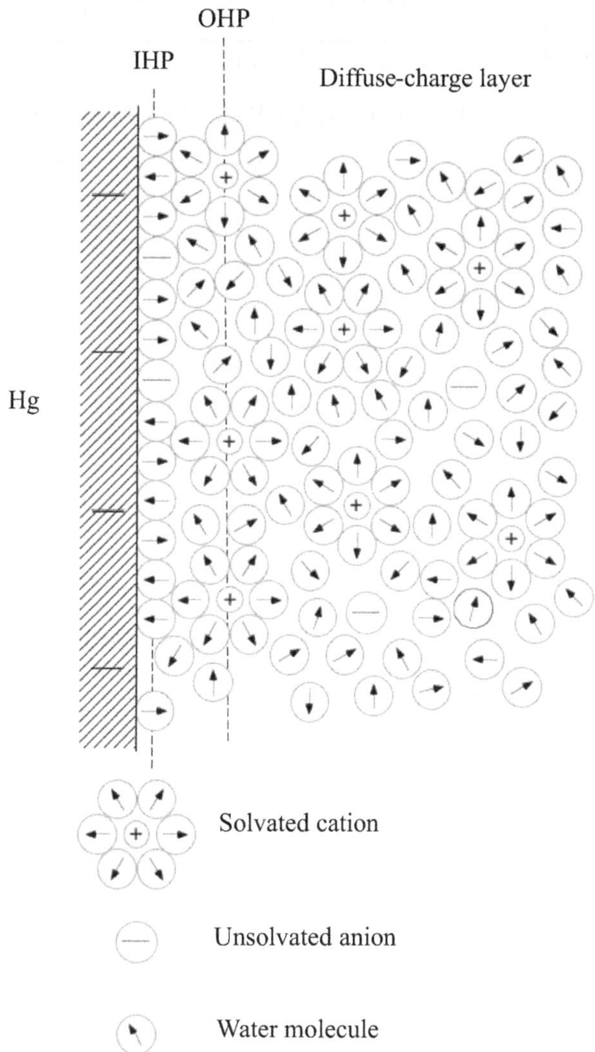

Fig. 1.7 Solvent adsorption model at the negatively charged mercury electrode [29]. In Fig. 1.7, the diffuse-charge layer is added to the model proposed by Bockris et al. [31]. Reprinted from [29], Copyright 1970, with permission from Plenum Press, New York

measured by several methods such as cantilever bending and piezoelectric detection (see Chap. 2). Figure 1.10 shows the surface stress versus potential (Δg vs. E) curve measured by a cantilever bending method at a potential scan rate of 5 mV s^{-1} in the potential region between −0.24 and 0.80 V (SHE) for an unreconstructed, (111)-textured Au (111) thin-film electrode in pH 3.0, 0.5 M $NaClO_4$ solution [33]. The surface stress versus potential (Δg vs. E) curve as well as the surface tension versus

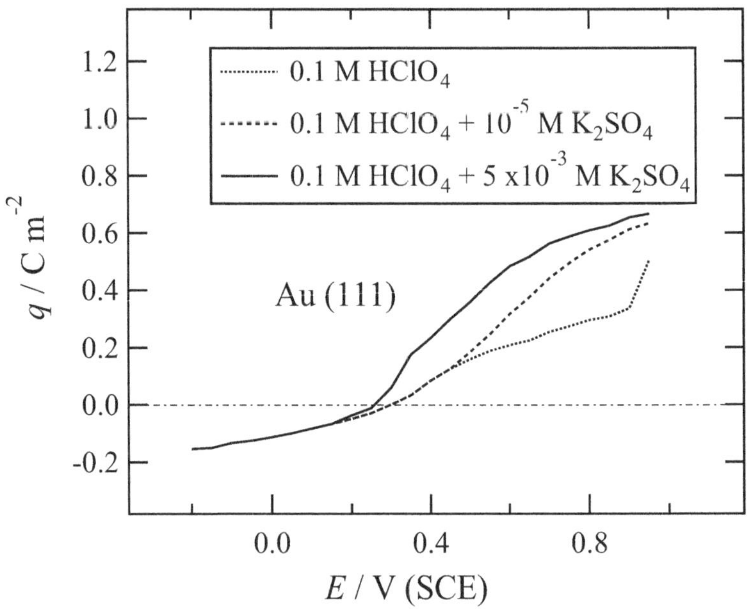

Fig. 1.8 Surface charge density versus potential (q vs. E) curves for an unreconstructed Au (111) electrode in 0.1 M $HClO_4$ solutions without and with 10^{-5} M, and 5×10^{-3} M K_2SO_4 [32]. Reprinted from [32], Copyright 1994, with permission from Elsevier

potential ($\Delta\gamma$ vs. E) curve is named "electrocapillary curve" for a solid electrode. In this book, the name "surface stress versus potential curve" is employed to distinguish from the electrocapillary curve ($\Delta\gamma$ vs. E) for a solid electrode.

The values of Δg in the ordinate of Fig. 1.10 are arbitrarily referred to zero at 0.80 V (SHE). The surface stress keeps almost constant in the potential range of −0.24 to 0.30 V (SHE), but it changes rapidly toward compressive direction ($\Delta g <$ 0) with increasing potential from 0.40 to 0.80 V (SHE). No clear maximum of the surface stress like an electrocapillary maximum (ecm) in Figs. 1.6 and 1.9 is observed in Fig. 1.10. As seen from the comparison in the same potential width between Figs. 1.9 and 1.10, the changes in surface stress are several times as much as the changes in surface tension. The similar results of the changes in surface stress for the unreconstructed Au (111) in pH 2, 0.1 M NaF and Na_2SO_4 solutions were reported by Vasiljevic et al. [34]. Schmickler and Leiva [18] calculated the contribution of the metal side to γ and g within the jellium model and found that the changes in g are much larger than those in γ, indicating that the surface concentration of electrons has a direct effect on the surface stress.

On the other hand, the maxima of the surface stress have been often observed in the Δg vs. E curves measured for a polycrystalline Au ribbon [35] or foil [36] electrode, and for a vacuum-deposited [37] or sputtered [38] Au thin-film electrode. Figure 1.11 shows the Δg vs. E curve measured in the potential region between −0.50 and 1.08 V (SHE) for a vacuum-deposited Au thin-film electrode in 0.1 M

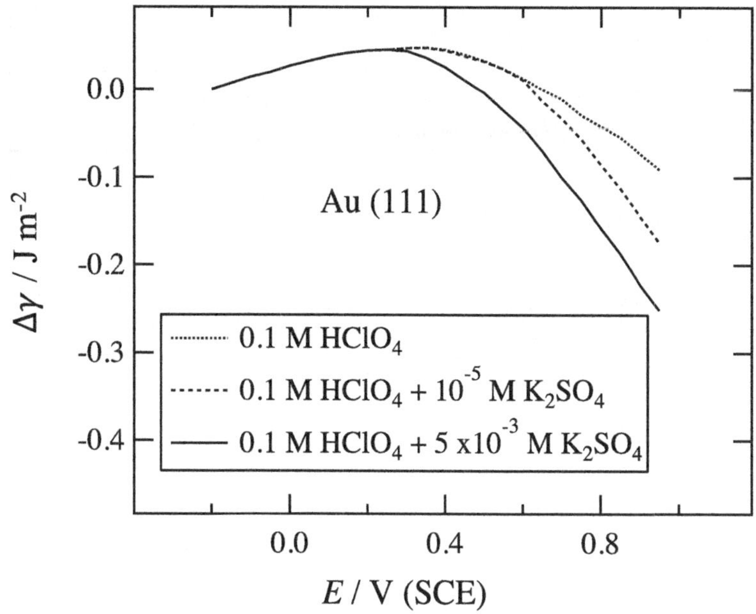

Fig. 1.9 Electrocapillary curves ($\Delta\gamma$ vs. E) obtained from Fig. 1.8 for the unreconstructed Au (111) electrode in 0.1 M $HClO_4$ solutions without and with 10^{-5} M, and 5×10^{-3} M K_2SO_4 [32]. Modified from [32], Copyright 1994, with permission from Elsevier

KCl solution [37]. The values of Δg are referred to zero at 1.08 V (SHE). The potential is shifted from 0.42 to −0.50 V (SHE), then up to 1.08 V and returned to 0.42 V (SHE). The maximum of the surface stress is observed at around 0 V in the potential scan from 0.42 V (SHE) toward negative direction. In the early studies of the electrocapillary phenomena for solid metal electrodes before 1998, there were some confusions for the use of two parameters: surface tension and surface stress. The terminology of surface tension γ or specific surface energy γ^s was often used in the case where the quantity measured experimentally is really the surface stress. Although the surface stress versus potential curve of a solid electrode has often a maximum, the surface stress versus potential curve is not equivalent to the surface tension versus potential curve.

The Gokhshtein equation of Eq. (1.119) indicates that the surface stress versus potential curve has no maximum if the following inequality holds in the entire potential region:

$$-q < \left(\frac{\partial q}{\partial \varepsilon}\right)_E . \tag{1.134}$$

On the other hand, in the case where the above inequality is reversed in some potential range, the surface stress versus potential curve has a maximum at a certain potential

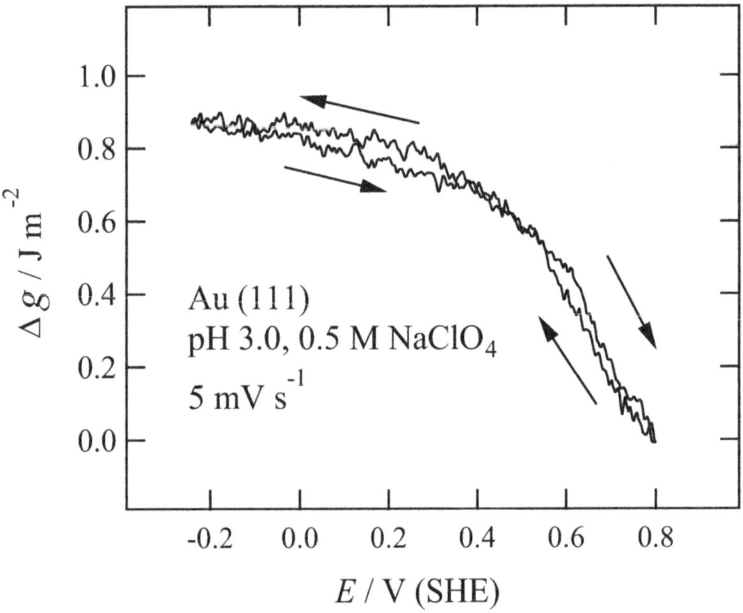

Fig. 1.10 Surface stress versus potential (Δg vs. E) curve measured by a cantilever bending method at a potential scan rate of 5 mV s^{-1} in the potential region between −0.24 and 0.80 V (SHE) for an unreconstructed, (111)-textured Au thin-film electrode in pH 3.0, 0.5 M $NaClO_4$ solution [33]. The values of Δg in the ordinate of Fig. 1.10 are arbitrarily referred to zero at 0.80 V (SHE). Reproduced from [33] with permission from The Electrochemical Society

corresponding to $-q = \left(\frac{\partial q}{\partial \varepsilon}\right)_E$. As pointed out by Valincius [25], the potential of the surface stress maximum is not equal to the potential of zero charge. If $\left(\frac{\partial q}{\partial \varepsilon}\right)_E$ has a positive value, the potential of the surface stress maximum should be more negative than the potential of zero charge E_{pzc}. Schmickler and Leiva [18] suggested the possibility for the surface stress passing through a maximum in the vicinity of E_{pzc} in the absence of contact or specific adsorption. The strain dependence of the surface charge density $\left(\frac{\partial q}{\partial \varepsilon}\right)_E$ is regarded as one of the important parameters for characterizing a solid electrode.

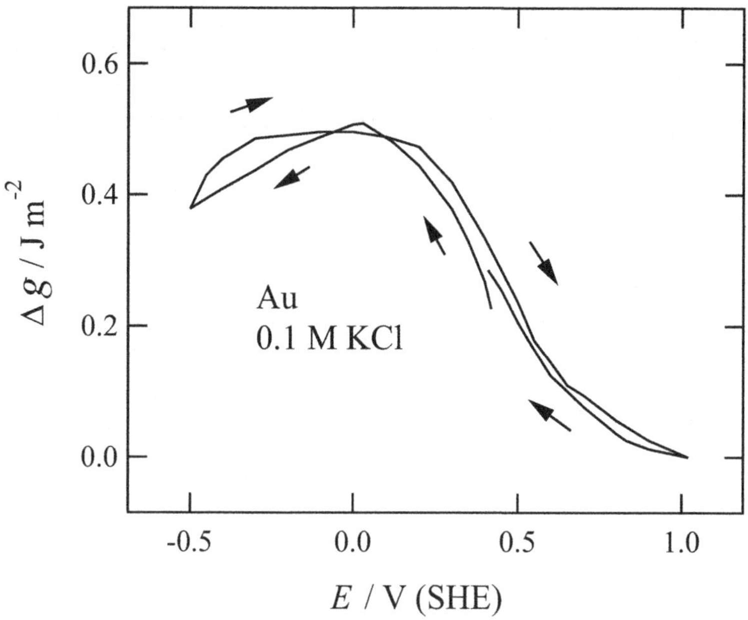

Fig. 1.11 Surface stress versus potential (Δg vs. E) curve measured in the potential region between 1.08 and −0.50 V (SHE) for a vacuum-deposited Au thin-film electrode in 0.1 M KCl solution [37]. The values of Δg are referred to zero at 1.08 V (SHE). The potential is shifted from 0.42 to −0.50 V, then up to 1.08 V and returned to 0.42 V (SHE). Reprinted from [37], Copyright 1971, with permission from Elsevier

Appendix 1: Derivation of the Tensor Equivalent of the Shuttleworth Equation

Let us follow the derivations made by Linford [11]. Consider the cube whose edges are unit length and parallel to x-, y-, and z-axes of a Cartesian reference frame. We perform two distinct operations on this cube as shown in Fig. 1.12 [10].

(1) Stretch the cube along the x-axis reversibly by an amount dx, maintaining the y edge at unit length, but allowing the z edge to vary its length. In this process, the flow of materials from the bulk by reducing the height of the crystal is permitted to form the extended surface. Let the work expended here denote by W_0. Next, let the stretched cube be separated (cleaved) along an xy plane, which requires work W_2:

$$W_2 = 2(\gamma + \Delta\gamma)(1 + dx), \tag{1.135}$$

where $\Delta\gamma$ is the variation in surface tension γ arising from the stretch dx. The factor of two arises due to the formation of two surfaces, and the term of $(1 + dx)$ is the area of an xy cross section of the stretched cube.

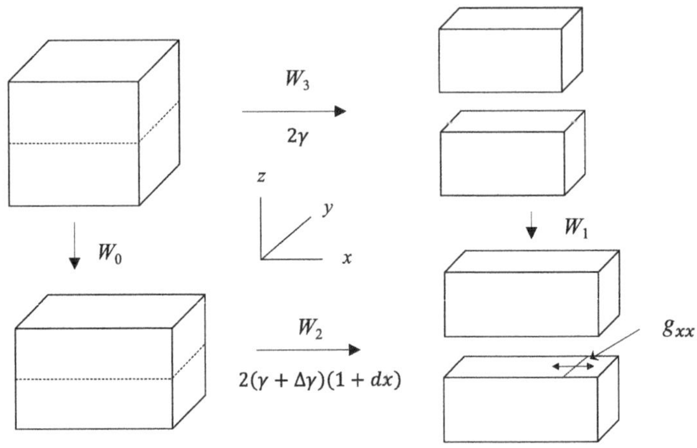

Fig. 1.12 Two distinct operations of stretching (W_0) followed by separating (W_2) or of separating (W_3) followed by stretching (W_1) on a cube [10]. The reversible works of $W_0 + W_2$ are equivalent to those of $W_1 + W_3$. Reprinted from [10] with permission of ASM International

(2) Now let the cube be returned to its original dimensions and then be separated (cleaved) along an xy plane into two halves. The work W_3 required to separate the cube into two halves is given by

$$W_3 = 2\gamma. \tag{1.136}$$

Furthermore, stretch each half in the x-direction by dx, keeping the y edge but the z edge not constant. Let this work denote by W_1. The work involved on the route (1), in which the cube is first stretched and then separated, is the same as that on the route (2), in which the cube is first separated and then stretched. As a result, the following relationship holds:

$$W_0 + W_2 = W_1 + W_3. \tag{1.137}$$

The substitution of Eqs. (1.135) and (1.136) into Eq. (1.137) leads to

$$W_0 + 2(\gamma + \Delta\gamma)(1 + dx) = W_1 + 2\gamma. \tag{1.138}$$

The difference ($W_1 - W_0$) in the work required in the stretching stage of the two routes is two times as much as the product of the component g_{xx} of a force (surface stress) in the newly formed surface and of the distance dx through which the force acts, that is,

$$W_1 - W_0 = 2g_{xx}dx. \tag{1.139}$$

It is noticed that the surface stresses of the unstretched two halves after the cleavage of the cube are balanced by elastic volume stresses in the interior of the two halves and thus are not included in Eq. (1.139) [10]. Since the strain $\Delta\varepsilon_{xx}$ is $\frac{dx}{1}$ in the unit cube, the substitution of Eqs. (1.139) into (1.138) leads to

$$2g_{xx}\Delta\varepsilon_{xx} = 2\gamma\Delta\varepsilon_{xx} + 2\Delta\gamma + 2\Delta\gamma\Delta\varepsilon_{xx}. \tag{1.140}$$

Because of $\Delta\gamma\Delta\varepsilon_{xx} \approx 0$ and $\frac{\Delta\gamma}{\Delta\varepsilon_{xx}} \approx \frac{d\gamma}{d\varepsilon_{xx}}$, Eq. (1.140) can be rewritten as

$$g_{xx} = \gamma + \frac{d\gamma}{d\varepsilon_{xx}}. \tag{1.141}$$

Consider the following two routes (3) and (4) analogous to those of (1) and (2) except that the stretching stages are replaced by shearing stages.

(3) The unit cube is first sheared, changing the shape, but not the area, of the *xy* plane cross section, which requires work W_0', and then separated into two halves. The work W_2' required for the separation is given by

$$W_2' = 2\gamma + 2\Delta\gamma. \tag{1.142}$$

(4) The unit cube is first separated into two halves along an *xy* plane, requiring work $W_3' = W_3 = 2\gamma$, and the subsequent shear of the two halves requires work W_1'. Since $(W_0' + W_2')$ is equal to $(W_1' + W_3')$, the following relationship holds:

$$W_1' - W_0' = 2\Delta\gamma. \tag{1.143}$$

The difference $W_1' - W_0'$ in the work required in the shearing stage of the two routes is two times as much as the product of the component g_{xy} of a force in the newly formed surface and of the distance dy through which the force acts. Since the strain $\Delta\varepsilon_{xy}$ is $\frac{dy}{1}$ in the unit cube, $W_1' - W_0'$ is given by

$$W_1' - W_0' = 2g_{xy}\Delta\varepsilon_{xy}, \tag{1.144}$$

and

$$g_{xy} = \frac{d\gamma}{d\varepsilon_{xy}}. \tag{1.145}$$

The combination of Eqs. (1.141) and (1.145) leads to the tensor equivalent of the Shuttleworth equation (see Eq. (1.71)): $g_{nm} = \gamma\delta_{nm} + \left(\frac{\partial\gamma}{\partial\epsilon_{nm}}\right)$. The g_{nm} quantities are the components of the surface stress tensor g, where

$$g = \begin{pmatrix} g_{xx} & g_{xy} \\ g_{yx} & g_{yy} \end{pmatrix}, \tag{1.146}$$

and for an isotropic substance or for a crystal surface with a threefold or greater axis of symmetry,

$$g = g\begin{pmatrix} 1 & 0 \\ 0 & 1 \end{pmatrix}, \tag{1.147}$$

in which g_{xy} and g_{yx} vanish, and the normal components of the tensor is equal to the tensor itself (see Eq. (1.69) or Eq. (1.80)).

Appendix 2: Calculation of the Magnitude of Surface Elastic Strain from the Curvature Change of Cantilever Bending

The following relationship between curvature radius of the cantilever bending R and surface stress g holds [21, 39] (see Eq. (2.14) in Sect. 2.3.1 of Chap. 2):

$$\frac{1}{R} = \frac{6(1 - \nu_s)}{E_s d_s^2} g,$$

where E_s, ν_s, and d_s are Young's modulus, Poisson's ratio, and thickness of the cantilever substrate, respectively. The surface strain ε is given by

$$\varepsilon = \frac{d_s}{2R}, \tag{1.148}$$

and a linear relationship between ε and g for small deformations is

$$\varepsilon = \frac{3(1 - \nu_s)g}{E_s d_s}. \tag{1.149}$$

Furthermore, the term $\left(\frac{\partial \varepsilon}{\partial E}\right)_{T,\mu_i}$ can be derived from Eq. (1.149):

$$\left(\frac{\partial \varepsilon}{\partial E}\right)_{T,\mu_i} = \frac{3(1 - \nu_s)}{E_s d_s}\left(\frac{\partial g}{\partial E}\right)_{T,\mu_i}. \tag{1.150}$$

In the experiments by Ibach et al. [22], the value of $\frac{3(1-\nu_s)}{E_s d_s} = 5.23 \times 10^{-8}$ m N^{-1} was obtained from the values of $d_s = 0.3$ mm, $E_s = 81.6$ GPa, and $\nu_s = 0.573$ for the Au (111) cantilever electrode. Very small value of $\varepsilon = 5.23 \times 10^{-8}$ for $g = 1.0$ J m^{-2} is calculated from Eq. (1.149). The value of $\left(\frac{\partial g}{\partial E}\right)_{T,\mu_i} \approx 1$Nm^{-1}V^{-1} was also obtained from the surface stress measurement of the Au (111) cantilever electrode [22]. Consequently, the value of $\left(\frac{\partial \varepsilon}{\partial E}\right)_{T,\mu_i} \approx 5 \times 10^{-8}$ V^{-1} is calculated from Eq. (1.150) [21].

References

1. Gibbs JW (1961) The scientific papers of J Willard Gibbs. vol 1: thermodynamics. Dover, New York, reprinted, pp 314–331
2. Guggenheim EA (1950) Thermodynamics: an advanced treatment for chemists and physics, 2nd edn. North-Holland Pub Co, Amsterdam, pp 35–37
3. Láng GG, Barbero CA (2012) Laser techniques for electrode processes. chap 2. Springer, Berlin
4. Defay R, Prigogine I with the collaboration of Bellemans A (1966) Surface tension and adsorption. chap 2. In: Everett DH (trans) Longmans, Green & Co Ltd., London, pp 21–31
5. Blakely JM (1973) Introduction to the properties of crystal surfaces. Pergamon Press, Oxford
6. Couchman PR, Jesser WA, Kuhlmann-Wilsdorf D, Hirth JP (1972) Surf Sci 33:429–436
7. Shuttleworth R (1950) Proc Phys Soc London A63:444–457
8. Herring C (1952) The use of classical macroscopic concepts in surface-energy problems, chap 1. In: Gomer R, Smith CS (eds) Structure and properties of solid surfaces. University of Chicago Press, Chicago, pp 5–18
9. Linford RG (1978) Chem Rev 78:81–95
10. Mullins WW (1963) Metal surfaces: structure, energetics and kinetics. American Society for Metals, Metals Park, Ohio
11. Linford RG (1973) Surface thermodynamics of solids. chap 1. In: Green M (ed) Solid state surface science. vol 2. Marcel Dekker, New York, pp 107–111
12. Couchman PR, Jesser WA (1973) Surf Sci 34:212–224
13. Couchman PR, Everett DH, Jesser WA (1975) J Colloid Interface Sci 52:410–411
14. Couchman PR, Everett DH (1976) J Electroanal Chem 67:382–386
15. Parsons R (1980) Thermodynamic methods for the study of interfacial regions in electrochemical systems. In: Bockris JO'M, Conway BE, Yeager E (eds) Comprehensive treatise of electrochemistry. vol 1: the double layer. Plenum Press, New York, pp 1–44
16. Bard AJ, Faulkner LR (1980) Electrochemical methods: fundamentals and applications. Wiley, New York, pp 491–493
17. Couchman PR, Davidson CR (1977) J Electroanal Chem 85:407–409
18. Schmickler W, Leiva E (1998) J Electroanal Chem 453:61–67
19. Needs RJ, Mansfield M (1989) J Phys: Condens Matter 1:7555–7563
20. Payne MC, Roberts N, Needs RJ, Needels M, Joannopoulos JD (1989) Surf Sci 211(212):1–20
21. Lipkowski J, Schmickler W, Kolb DM, Parsons R (1998) J Electroanal Chem 452:193–197
22. Ibach H, Bach CE, Giesen M (1997) Grossmann. Surf Sci 375:107–119
23. Shi Z, Lipkowski J, Gamboa M, Zelenay P, Wieckowski A (1994) J Electroanal Chem 366:317–326
24. Vasiljevic N, Trimble T, Dimitrov N, Sieradzki K (2004) Langmuir 230:6639–6643
25. Valincius G (1999) J Electroanal Chem 478:40–49
26. Proost J (2005) J Solid State Electrochem 9:660–664
27. Gokhshtein AY (1976) Surface tension of solids and adsorption. Nauka, Moscow
28. Damaskin BB, Petrii OA (2002) Specific adsorption. In: Gileadi E, Urbakh M (eds) Encyclopedia of electrochemistry. vol 1: thermodynamics and electrified interfaces. Wiley-VCH, Weinheim
29. Bockris JO'M, Reddy AKN (1977) Modern electrochemistry. vol 2. Plenum Press, New York
30. Grahame DC (1947) Chem Rev 41:441–501
31. Bockris JO'M, Devanathan MAV, Muller K (1963) Proc Roy Soc London A274:55–79
32. Shi Z, Lipkowski J (1994) J Electroanal Chem 364:289–294
33. Seo M, Yamazaki M (2004) J Electrochem Soc 151:E276–E281
34. Vasiljevic N, Trimble T, Dimitrov N, Sieradzki K (2004) Langmuir 20:6639–6643
35. Lin KF, Beck TR (1976) J Electrochem Soc 123:1145–1151
36. Seo M, Jiang XC, Sato N (1987) J Electrochem Soc 134:3094–3098
37. Fredlein RA, Damjanovic A, Bockris JO'M (1971) Surf Sci 25:261–264
38. Láng GG, Heusler KE (1995) J Electroanal Chem 391:169–179
39. Ibach H (1997) Surf Sci Rep 29:193–263

Chapter 2
Methods for Investigating Electro-Chemo-Mechanical Properties of Solid Electrode Surfaces

Abstract Several methods powerful for investigating the electro-chemo-mechanical properties of solid electrode surfaces have been introduced and discussed. A piezoelectric response method can detect sensitively the differential surface stress. The principle and feature of the piezoelectric response method are explained, and its application results of a polycrystalline gold electrode in sulfuric acid solution are discussed. A cantilever bending method is capable of measuring the changes in surface stress from the curvature of a cantilever electrode. The principle and feature of the curvature measurement by a laser beam for the cantilever bending method are explained, and the issues for the curvature measurement are discussed. Furthermore, new techniques for measuring the change in surface stress due to the change in surface charge or the change in potential due to the elastic deformation of solid electrodes and for detecting the change in surface elastic strain of nano-porous metal electrodes are introduced and their principles and features are explained.

Keywords Surface stress measurement · Cantilever bending · Piezoelectric detection · Elastic deformation · Dilatometry

2.1 Introduction

The change in potential or in surface charge of a solid electrode induces the change in surface stress, while the elastic deformation of a solid electrode induces the change in potential. The measurements of the potential- or surface charge-induced surface stress change and of the elastic deformation-induced potential change are indispensable for investigating the electro-chemo-mechanical properties of solid electrode surfaces. The principles and features of various methods developed so far for the measurement of the surface stress change have been described and discussed in detail by Láng [1]. Particularly, piezoelectric detection and cantilever bending methods have been conveniently employed for the measurement of the changes in surface stress for noble metal electrodes such as Au and Pt. In addition, new methods for detecting sensitively the surface charge-induced surface stress change and the elastic deformation-induced

M. Seo, *Electro-Chemo-Mechanical Properties of Solid Electrode Surfaces*,
https://doi.org/10.1007/978-981-15-7277-7_2

potential change have been recently developed. The present chapter deals with the principles and features of several methods powerful for investigating the electro-chemo-mechanical properties of solid electrode surfaces.

2.2 Piezoelectric Detection of Differential Surface Stress

Gokhshtein [2–4] is the pioneer to measure the differential surface stress of a solid electrode by using a device consisting of mechanically connected piezoelectric plate and over-meniscus working metal electrode. Figure 2.1 shows the experimental setup of the device for measuring the differential surface stress (simultaneously at two

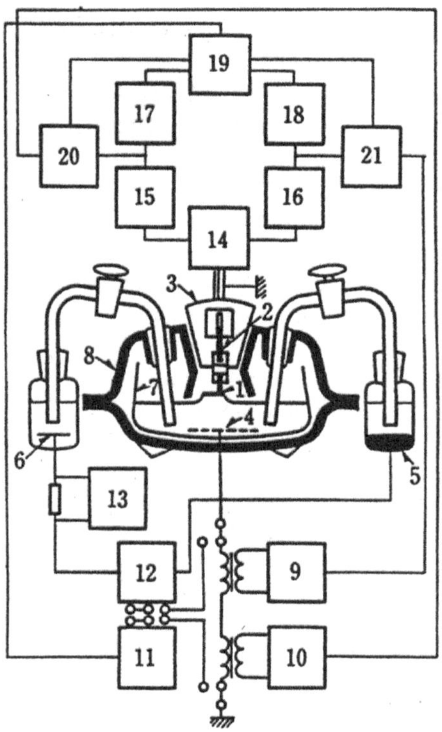

Fig. 2.1 Experimental setup of the device for measuring the differential surface stress (simultaneously at two frequencies) of a solid electrode designed by Gokhshtein [2, 3]. 1—Over-menisucus metal electrode; 2—Piezoelectric element; 3—Screen; 4—Auxiliary electrode; 5—Reference electrode; 6—Polarizing electrode; 7—Quartz vessel; 8—Glass cell; 9 and 10—Sinusoidal voltage generators; 11—Linear voltage generator; 12—Compensator; 13—Current amplifier; 14—Broadband pre-amplifier: 15 and 16—Selective amplifiers; 17 and 18—Rectifiers; 19—Multichannel oscillographic recorder; 20 and 21—Blocks for phase measurements. Reprinted from [3], Copyright 1970, with permission from Elsevier

frequencies) of a solid electrode designed by Gokhshtein [2, 3]. A small alternating voltage superimposed on an applied DC potential of the metal electrode induces an oscillatory variation of the surface stress, which leads to a mechanical oscillation in the entire electrode-piezoelectric element system. The potential modulation in the range of 0.5–20 kHz is chosen to induce the mechanical resonances of the whole system in which the amplitude of oscillation of the piezoelectric element becomes very large compared to non-resonance oscillations. The piezoelectric element converts the mechanical oscillations to alternating electric signals, which can be sensitively detected by using an oscilloscope, a lock-in amplifier, or a frequency response analyzer.

The oscillatory variation of the surface stress can also be induced by the application of a small alternating current at a high frequency, in which the amplitude of the surface charge density is kept constant. The amplitudes of the piezoelectric signals generated by the potential and current modulations are proportional to the absolute values $\left(\left|\frac{\partial g}{\partial E}\right| \text{and} \left|\frac{\partial g}{\partial q}\right|\right)$ of the differentials of surface stress g with respect to electrode potential E and with respect to surface charge density q, which are, respectively, named "E-estance" and "q-estance" by Gokhshtein [2–4]. Furthermore, the phase angles of the piezoelectric signals contain information of the changes in sign of $\frac{\partial g}{\partial E}$ and $\frac{\partial g}{\partial q}$. The following relationship between $\frac{\partial g}{\partial E}$ and $\frac{\partial g}{\partial q}$ holds:

$$\frac{\partial g}{\partial E} = \frac{\partial g}{\partial q}\frac{\partial q}{\partial E} = c\frac{\partial g}{\partial q}, \tag{2.1}$$

where c is the differential capacity of the electrode. The $\frac{\partial g}{\partial E}$ versus E or $\frac{\partial g}{\partial q}$ versus E curve has a minimum amplitude (nearly zero) of the piezoelectric signal with a phase angle change of 180° at a certain potential, which corresponds to the surface stress maximum. The potential at the surface stress maximum for a solid electrode is not equal to the potential of zero charge $E_{pzc}(q = 0)$ since $q = -\frac{\partial q}{\partial \varepsilon} \neq 0$ holds at the surface stress maximum as manifested by the Gokhshtein equation (see Eq. (1.119) in Sect. 1.8 of Chap. 1). Valincius [5, 6] measured the q-estance $\left(\frac{\partial g}{\partial q} \text{ vs. } E\right)$ curves for a polycrystalline Au in $HClO_4$ solutions with different concentrations and found the negative shift of the potential at the surface stress maximum for the Au electrode with respect to E_{pzc} determined from the differential capacity measurements. This means that $\frac{\partial q}{\partial \varepsilon} > 0$ holds near E_{pzc} for the polycrystalline Au electrode.

The device for the piezoelectric detection of differential surface stress was modified by several researchers [7–11]. Figure 2.2 shows the piezoelectric electrode modified by Seo et al. [9, 10], in which a metal foil plate (working electrode) is attached via a thin polyimide film for electrical isolation to a piezoelectric ceramic plate with strain gauge cement and then doubly coated with epoxy cement and silicon sealant for electrical isolation from solution in the back side. The modified device combined with piezoelectric element and working electrode can be used in electrolyte solutions. Seo et al. [9, 10] by using the piezoelectric electrode in Fig. 2.2 measured the differential surface stress induced by a potential modulation for polycrystalline Pt

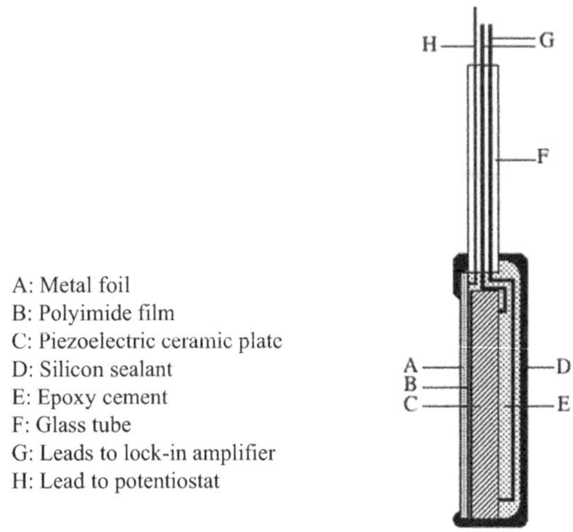

Fig. 2.2 Device modified by Seo et al. [9, 10], in which a metal foil plate (working electrode) is attached via a thin polyimide film for electrical isolation to a piezoelectric ceramic plate with strain gauge cement and doubly coated with epoxy cement and silicon sealant for electrical isolation from solution in the back side

and Au foil electrodes in sulfate solutions of different pH values. Figure 2.3 shows the block diagram of the apparatus used in experiments [9]. A potentiostat connected with a function generator is used to supply a linear potential scan (20–100 mV s^{-1}) to the working electrode. A sinusoidal signal of 5 mV at 150–50 Hz supplied from an oscillator is superimposed on the linear potential scan.

This small superimposed alternating signal d*E* to the electrode induces an alternating change in surface stress *dg* of the electrode surface, which is directly transmitted to a piezoelectric ceramic plate. The signal from the piezoelectric ceramic plate

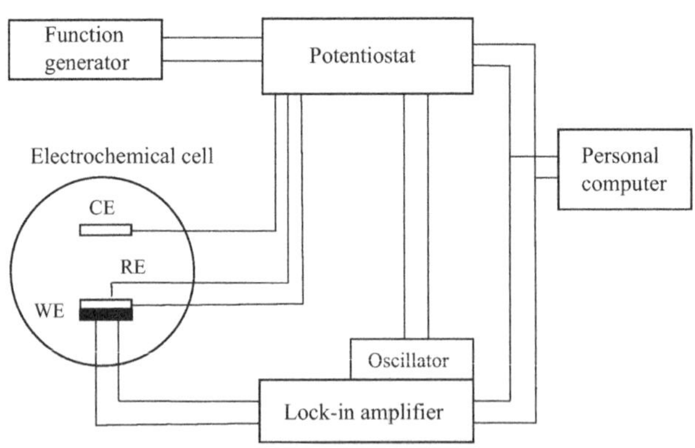

Fig. 2.3 Block diagram of the apparatus used for the piezoelectric detection of differential surface stress [9, 10]. Reproduced from [9] with permission from The Electrochemical Society

is synchronously detected at the same frequency as that of superimposed potential modulation by using a lock-in amplifier. A simultaneous recording or data acquisition with a personal computer is performed for measuring the cyclic voltammogram and piezoelectric signal curve (amplitude $|A|$ vs. phase angle φ) in the course of cyclic potential scan. In principle, the amplitude $|A|$ is proportional to $\left|\frac{\partial g}{\partial E}\right|$ and the phase angle φ involves the component of the change in sign of $\frac{\partial g}{\partial E}$. The components resulting from the instruments and the mechanical properties of the electrode system are included in φ. Nevertheless, if the contribution of the additional components to φ is kept constant during experiments, the change in sign of $\frac{\partial g}{\partial E}$ can be evaluated from the relative change of φ by 180°.

Figure 2.4 shows (a) the cyclic voltammogram and (b) the corresponding piezoelectric signal curve ($|A|$ vs. E and φ vs. E) for a polycrystalline Au foil electrode in 0.5 M H_2SO_4 solution [10]. The cyclic voltammetry was performed at a potential scan rate of 33.3 mV s^{-1}, and simultaneously a sinusoidal signal of 5 mV at 200 Hz was superimposed on the linear potential scan. The minima of $|A|$ with the change of φ by 180° are observed at about—0.05 V (SHE) in the cathodic potential scan and at about 0.15 V (SHE) in the anodic potential scan, respectively, which correspond to the maxima of surface stress. The potentials of the surface stress maximum E_{max}, despite the hysteresis of about 0.2 V in the anodic and cathodic potential scans, are close to those obtained by scrape [12] and ribbon extension [13] methods. The value of E_{max} is independent of pH in acidic and neutral sulfate solutions, while at pH higher than 8, E_{max} shifts toward negative direction, indicating that adsorption of OH^- ions on Au proceeds in alkaline solutions [10]. The relationship between Δg and $|A|$ can be given by

$$\Delta g = \int\left|\frac{\partial g}{\partial E}\right|\mathrm{d}E = K\int|A|\mathrm{d}E, \tag{2.2}$$

where K is the conversion coefficient from $|A|$ to $\left|\frac{\partial g}{\partial E}\right|$. It is difficult to determine K accurately because the mechanical coupling of the piezoelectric electrode is complicated.

Figure 2.5 shows the Δg versus E curve obtained from the graphical integration of $|A|$ with respect to E in the anodic potential scan in Fig. 2.4b by taking the sign-reversal of $\frac{dg}{\mathrm{d}E}$ at about 0.2 V (SHE) into account. Seo et al. [10] neglected the second term of the Gokhshtein equation, i.e., $\frac{\partial q}{\partial \varepsilon}$, and determined $K = 0.35$ C m^{-2} per μV from the linear relationship between $|A|$ and q in the potential range of $E > E_{max}$. However, as pointed out by Valincius [5, 6], the term of $\frac{\partial q}{\partial \varepsilon}$ as compared to q cannot be neglected and the potential of the surface stress maximum E_{max} is not equal to the potential of zero charge E_{pzc}. Valincius [5] proposed two procedures to calibrate the conversion coefficient K. Figure 2.6 illustrates the calibration procedures of K from two-coupled $|A|$ vs. E and q vs. E (a, b and a', b) curves in the vicinity of E_{pzc} for two electrolyte solutions with different concentrations. In the limiting case where $\frac{\partial q}{\partial \varepsilon}$

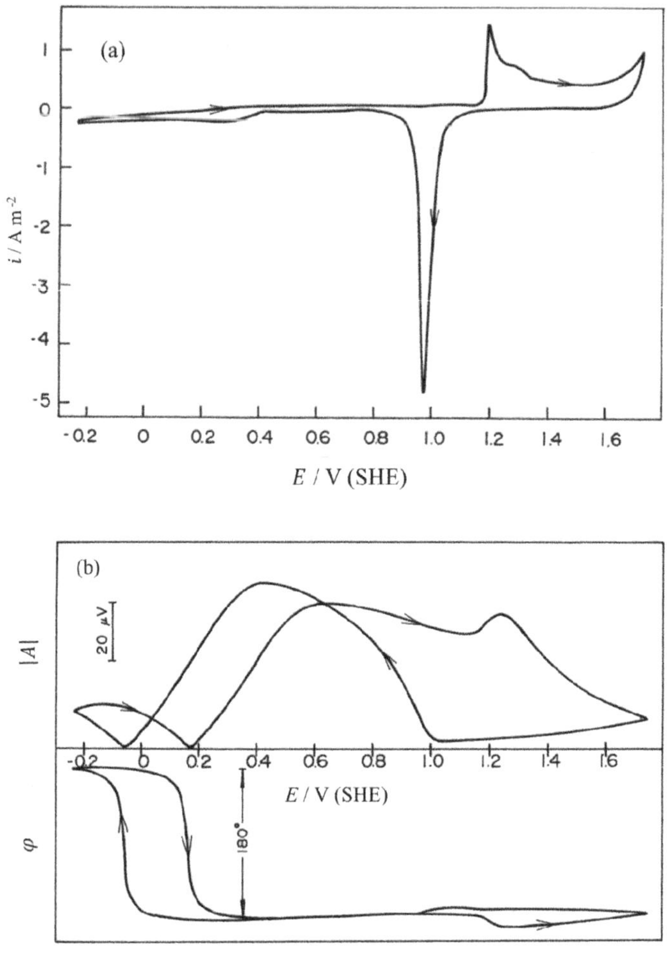

Fig. 2.4 **a** Cyclic voltammogram and **b** the corresponding piezoelectric signal curve ($|A|$ vs. E and φ vs. E) for a polycrystalline Au foil electrode in 0.5 M H_2SO_4 solution [10]. Reproduced from [10] with permission from The Electrochemical Society

keeps constant in the vicinity of E_{pzc}, K can be calibrated from the coupled a and b curves in Fig. 2.6.

The following relationships hold from the Gokhshtein equation (see Eq (1.121) in Sect. 1.8 of Chap. 1):

$$-\frac{\partial g_1}{\partial E} = q_1 + \frac{\partial q_1}{\partial \varepsilon}, \tag{2.3}$$

and

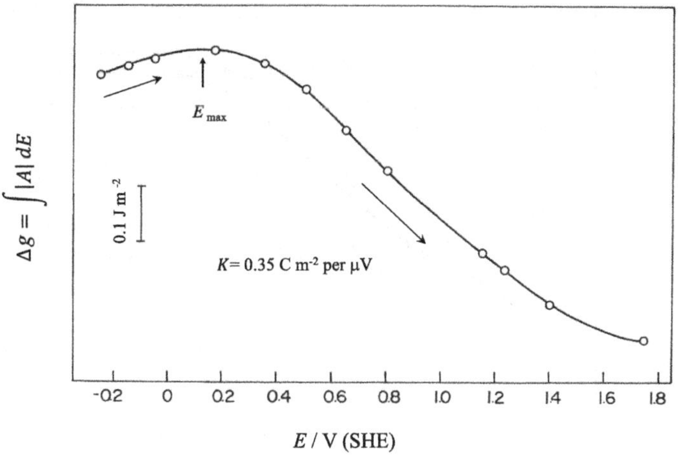

Fig. 2.5 Δg versus E curve obtained from the graphical integration of $|A|$ with respect to E in the anodic potential scan in Fig. 2.4b by taking the sign-reversal of $\frac{dg}{dE}$ at about 0.2 V (SHE) into account [10]. The conversion coefficient $K = 0.35$ C m^{-2} per μV is determined from the linear relation between $|A|$ and q in the potential range of $E > E_{max}$ in Fig. 2.4b. Reproduced from [10] with permission from The Electrochemical Society

$$-\frac{\partial g_2}{\partial E} = q_2 + \frac{\partial q_2}{\partial \varepsilon}, \tag{2.4}$$

where the values of q_1 and q_2 are taken within a small potential interval of 30–50 mV in the vicinity of E_{pzc}. Assuming that $\frac{\partial q_1}{\partial \varepsilon}$ is equal to $\frac{\partial q_2}{\partial \varepsilon}$ in the small potential interval, K is given by

$$K = \frac{-\frac{\partial g_2}{\partial E} + \frac{\partial g_1}{\partial E}}{|A_2| - |A_1|} = \frac{q_2 - q_1}{|A_2| - |A_1|}. \tag{2.5}$$

In Eq. (2.5), the minus sign of the piezoelectric signals is taken at $E > E_{max}$. Equation (2.5) indicates that K can be experimentally obtained since all parameters are known from the measured a and b curves. The calibration of K in Fig. 2.5 is based on Eq. (2.5). On the other hand, in the case where $\frac{\partial q}{\partial \varepsilon}$ depends on potential in the vicinity of E_{pzc}, the calibration may be achieved by using two-coupled piezoelectric signal and surface charge density data (a, b and a', b' in Fig. 2.6), obtained in solutions with two different concentrations at least. The following equations as well as Eqs. (2.3) and (2.4) hold:

$$-\frac{\partial g_3}{\partial E} = q_3 + \frac{\partial q_3}{\partial \varepsilon}, \tag{2.6}$$

and

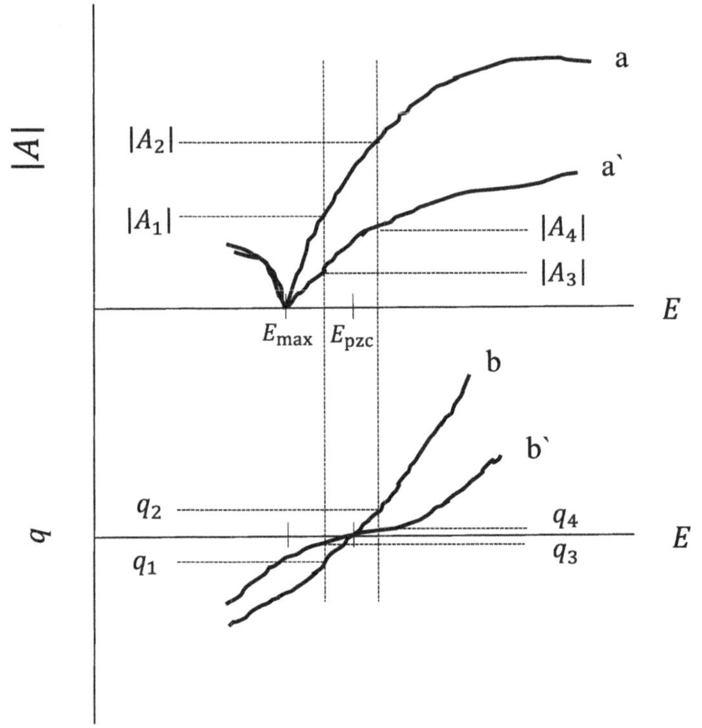

Fig. 2.6 Schematic two-coupled $|A|$ vs. E and q vs. E (a, b and a', b') curves in the vicinity of E_{pzc} for two electrolyte solutions with different concentrations for calibration of the conversion coefficient K [5]. Modified with the permission from [5], Copyright 1998, American Chemical Society

$$-\frac{\partial g_4}{\partial E} = q_4 + \frac{\partial q_4}{\partial \varepsilon}. \tag{2.7}$$

Then, K may be written in place of Eq. (2.5):

$$K = \frac{\left(-\frac{\partial g_2}{\partial E} + \frac{\partial g_1}{\partial E}\right) - \left(-\frac{\partial g_4}{\partial E} + \frac{\partial g_3}{\partial E}\right)}{(|A_2| - |A_1|) - (|A_4| - |A_3|)}. \tag{2.8}$$

There are unknown parameters to determine K from Eq. (2.8). The term of $\Delta\left(\frac{\partial q}{\partial \varepsilon}\right)$ is elastic by nature and is not influenced by a diffuse part of the electric double layer which is purely plastic because it is located in liquid phase. In addition, the variation of the electrolyte concentration (at least below 0.1 M) does not affect the structure of the compact (inner Helmholtz) part of the electric double layer near E_{pzc}. This leads to the reasonable assumption that $\left(\frac{\partial q_2}{\partial \varepsilon} - \frac{\partial q_1}{\partial \varepsilon}\right)$ and $\left(\frac{\partial q_4}{\partial \varepsilon} - \frac{\partial q_3}{\partial \varepsilon}\right)$ are independent of concentration; i.e., $\left(\frac{\partial q_2}{\partial \varepsilon} - \frac{\partial q_1}{\partial \varepsilon}\right)$ is equal to $\left(\frac{\partial q_4}{\partial \varepsilon} - \frac{\partial q_3}{\partial \varepsilon}\right)$ [5]. Consequently, K is eventually given by:

$$K = \frac{(q_2 - q_1) - (q_4 - q_3)}{(|A_2| - |A_1|) - (|A_4| - |A_3|)}. \tag{2.9}$$

The value of K can be directly obtained from the respective values of $|A|$ and q in the measured a, a', b, and b' curves (Fig. 2.6).

One of the important results obtained by the piezoelectric detection of differential surface stress is that another minimum of $|A|$ with a change of φ by 180° (i.e., sign-reversal of differential surface stress) emerges often in the high potential region where a monolayer of surface oxide such as PtO and PdO [2, 9, 14–16] is formed. It is presumed from the Gokhshtein equation that the sign-reversal of the differential surface stress takes place when the sign of $\frac{\partial q}{\partial \varepsilon}$ changes from plus to minus so as to satisfy the inequality of $-q > \frac{\partial q}{\partial \varepsilon}$ after passing through $-q = \frac{\partial q}{\partial \varepsilon}$ in the course of the monolayer formation of surface oxide. The piezoelectric electrode has been further improved by Seo and Ueno [17] as shown in Fig. 2.7. The improved working electrode consists of a polycrystalline metal foil (20 mm in diameter × 20 μm) with a piezoelectric ceramic disk (20 mm × 1 mm) attached to its back side by a strain

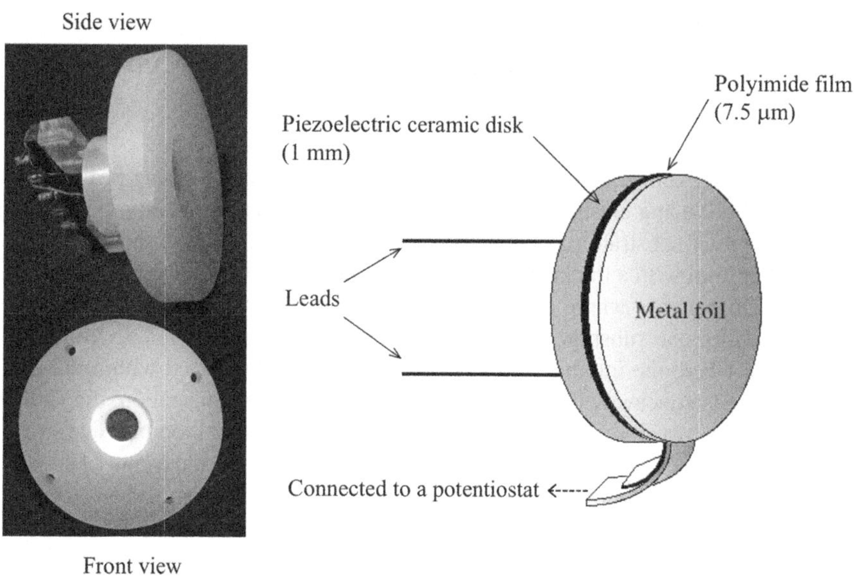

Fig. 2.7 Structure of the piezoelectric electrode improved by Seo and Ueno [17]. The improved working electrode consists of a polycrystalline metal foil (20 mm in diameter × 20 μm) with a piezoelectric ceramic disk (20 mm × 1 mm) attached to its back side by a strain gauge cement via a thin polyimide film (7.5 μm) for electric isolation. The working electrode is encapsulated into a Teflon holder

gauge cement via a thin polyimide film (7.5 μm) for electric isolation. The working electrode is encapsulated into a Teflon holder. The electrochemical cell is composed of a Teflon cylinder into which the working electrode assembly is clamped from one end.

This working electrode assembly has been employed for differential surface stress measurements of a polycrystalline Au foil electrode in 1 M $NaClO_4$ solutions containing 10^{-4}–10^{-2} M KI [17]. The surface stress maximum of the Au electrode in 1 M $NaClO_4$ solution without KI appeared at $E_{max} = 0.13$ and 0.02 V (SHE) in the anodic and cathodic potential scans (20 mV s^{-1}), respectively. On the other hand, the mean value of E_{max} shifted toward negative direction by 0.64 V due to an addition of 3×10^{-3} M KI in the solution, indicating a strong contact adsorption of iodide ions. However, Seo et al. in this paper [17] as well as other their papers [9, 10, 14–16] neglected the term of $\frac{\partial q}{\partial \varepsilon}$ as compared to the surface charge density q and regarded as $E_{max} = E_{pzc}$. Afterward, Seo et al. [18, 19] investigated the relationship between $\frac{\partial g}{\partial E}$ and q by using a cantilever bending method (see Sect. 2.3 of this chapter) and found a significant deviation of $\frac{\partial g}{\partial E}$ from q at potentials more positive than E_{pzc}.

2.3 Cantilever Bending Method for Measurement of Changes in Surface Stress

2.3.1 Relationship between Surface Stress and Curvature of Cantilever

Surface stress g can be calculated from a cantilever bending (bending beam, wafer curvature, etc.) using elasticity theory. Figure 2.8 shows schematically the bending of a rectangular substrate (the both edges are not clamped) with a thickness of d_s when a thin film with a thickness of d_f is uniformly deposited on the top side of the substrate. In the case where the internal compressive stress ($\sigma_f < 0$) is developed within the film, the substrate is convexly bended as shown in Fig. 2.8a. Similarly, in the case where the internal tensile stress ($\sigma_f > 0$) is developed within the film, the substrate is concavely bended as shown in Fig. 2.8b. By convention, the sign of curvature κ or curvature radius R is minus for convex bending (compressive stress) and plus for concave bending (tensile stress) in accordance with the sign of stress.

In the mechanical equilibrium of the bending, the compressive forces or tensile forces developed within the film are balanced by the tensile or compressive forces in the substrate; i.e., the net force F vanishes and simultaneously the net bending moment M also vanishes on the film-substrate cross section. Therefore, the following relationships hold [1]:

$$F = \int \sigma \mathrm{d}A = 0, \tag{2.10}$$

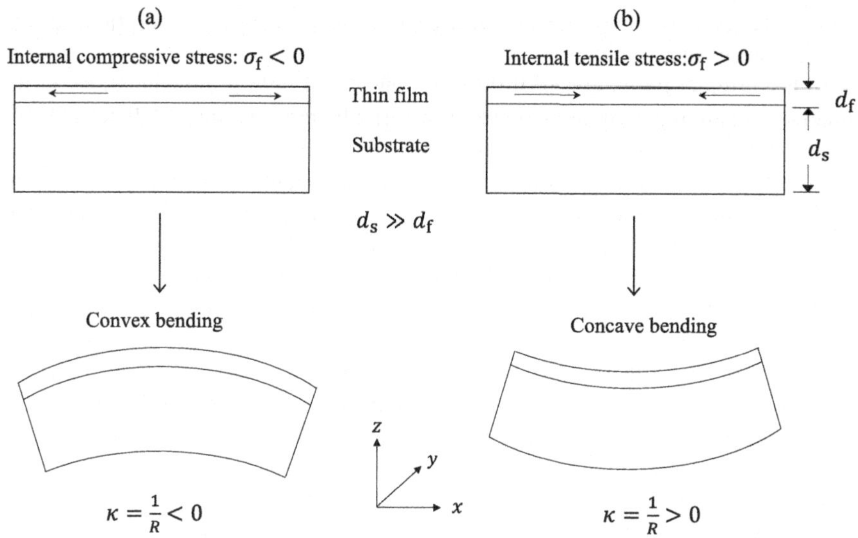

Fig. 2.8 Scheme for the bending of a rectangular substrate (the both edges are not clamped) with a thickness of d_s when a thin film with a thickness of d_f is uniformly deposited on the top side of the substrate: **a** the substrate is convexly bended in the case where the internal compressive stress ($\sigma_f < 0$) is developed within the film, and **b** the substrate is concavely bended in the case where the internal tensile stress ($\sigma_f > 0$) is developed within the film. The curvature of the substrate κ is reciprocal of the curvature radius R

and

$$M = \int \sigma z \mathrm{d}A = 0, \tag{2.11}$$

where A is the sectional area (z–y plane in Fig. 2.8), and z is the distance from the neutral plane (where an elastic strain becomes zero) in the substrate. Elasticity theory was first applied by Stoney [20] to derive the relationship between surface stress and curvature of cantilever. Stoney considered a "thin steel rule" with a thin nickel layer with a thickness d_f deposited on steel. Assuming that the thickness of the rule $d_s(\gg d_f)$ is negligibly small as compared to the radius of curvature R, the following relationship was derived [1, 20]:

$$g = \sigma_f d_f = \frac{E_s \kappa d_s^2}{6} = \frac{E_s d_s^2}{6R}, \tag{2.12}$$

where σ_f is the stress in the film, κ is the curvature (equal to reciprocal of R), and E_s is Young's modulus of the substrate. The unit of $\sigma_f d_f$ in Eq. (2.12) is J m^{-2}, which corresponds to the unit of surface stress g. Equation (2.12) is also valid in the case where the uppermost surface (free from a thin film) on the top side of the substrate is subjected to compressive or tensile surface stress due to charging, adsorption, surface reconstruction, etc.

In his original derivation [20], Stoney considered only a uniaxial stress in the direction of bending, a situation of which is normally not encountered. Ibach [21–23] modified the Stoney's original equation by taking into account the biaxial nature of the stress and the geometry of the specimen under different boundary conditions. The modified Stoney's equation can be derived from the balancing of bending moments or torques along the principal axes in the solid, i.e., the vanishing of net bending moment [1, 22] (see Eq. (2.11)). Alternatively, the modified Stoney's equation can be also derived from a minimum of the total elastic energy per unit area in the sample plate at a particular curvature [21, 23]. Ibach [21] derived the modified equation from the minimization of the total elastic energy (including surface and bulk) based on elasticity theory under two boundary conditions: one is that the bending is allowed only in one direction (i.e., uniaxial bending), and the other one is that the bending is free in both principal directions (i.e., biaxial bending).

For simplicity, it is considered that a cubic crystal plate is oriented to such that the surfaces are (100) surfaces and that the sides of the rectangular shaped plate coincide with the <100> directions [21]. For a uniaxial bending, the following modified Stoney's equation was eventually derived from the minimization of the total elastic energy based on elasticity theory [21]:

$$g_{11} = \frac{E_s \kappa_1 d_s^2}{6\left(1 - \nu_s^2\right)} = \frac{E_s d_s^2}{6R\left(1 - \nu_s^2\right)}, \tag{2.13}$$

where g_{11} and κ_1 are the surface stress (product $\sigma_f d_f$) and the curvature for bending of one principal direction, respectively. The modified Stoney's equation for a biaxial bending was also derived [21]:

$$g = \frac{E_s \kappa d_s^2}{6(1 - \nu_s)} = \frac{E_s d_s^2}{6R(1 - \nu_s)}. \tag{2.14}$$

The modified Stoney's equations for uniaxial and biaxial bending contain the terms of $\left(1 - \nu_s^2\right)$ and $(1 - \nu_s)$, respectively, in the denominator as compared to the original one [see Eq. (2.12)]. Ibach [21] confirmed that Eqs. (2.13) and (2.14) are also valid for the cubic crystal plate, the surface of which is oriented to (111) surface. Equations (2.13) and (2.14) are named "generalized Stoney's equations." Moreover, a finite element analysis of cantilever bending based on elasticity theory [23] indicated that the generalized Stoney's equations are applicable for general surfaces of all crystals.

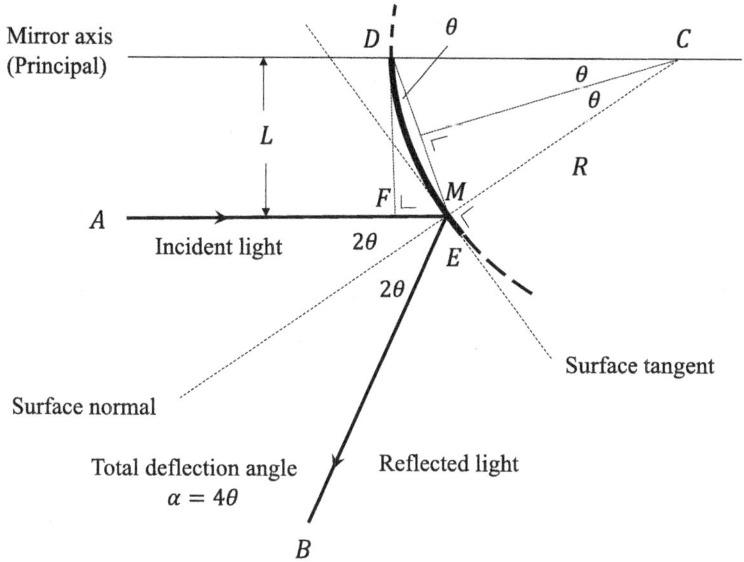

Fig. 2.9 Scheme for deriving the relationship between the total deflection angle α and the curvature radius R or curvature κ for the cantilever bending [1, 24]. The incident light is parallel to the mirror axis. Modified from [24], Copyright 2010, with permission from AIP Publishing

2.3.2 *Optical Detection of Curvature*

Popular method for measuring the curvature of cantilever plate is a laser beam optical detection, which is based on the reflection of the incident light at the cantilever surface (as a mirror). Figure 2.9 shows the scheme for deriving the relationship between the deflection angle α and the curvature radius R or curvature κ for bending of the cantilever [1, 24]. The incident light is reflected at a point M on the segment DE (depicted as a thick solid curve in Fig. 2.9) of a convex spherical or cylindrical (mirror) surface, which corresponds to the cantilever. The angle of incidence 2θ is measured from a normal line drawn to the curved surface at the reflection point M. The angle of reflection is equal to the angle of incidence, since the law of reflection holds at each point of the curved surface. Consequently, the total angle α of deflection of the light is

$$\alpha = 4\theta. \tag{2.15}$$

The normal line drawn to the curved surface at M intersects at a point C (i.e., center of the curvature) with a normal line (i.e., mirror (principal) axis) drawn to the same curved surface at a point D. The mirror axis is parallel to the incident light. The length MC or DC corresponds to the curvature radius R (reciprocal of the curvature κ). The bending of the cantilever within elastic limit is very small, i.e., $2\theta \approx 0$, so that the distance L defined as a length DF in Fig. 2.9 can be derived [1, 24]:

$$L = 2R \sin\theta \cos\theta \approx 2R\theta, \tag{2.16}$$

and

$$\frac{1}{R} = \kappa \approx \frac{2\theta}{L} = \frac{\alpha}{2L}. \tag{2.17}$$

As the cantilever bends, the total deflection angle α increases to bring the changes in the spot of the reflected light on a detector plane, from which the changes in $\frac{1}{R}$ or κ are calculated by using Eq. (2.17).

Figure 2.10 shows the optical configuration of the cantilever bending setup in vacuum or air [1, 24]. The deflection of the laser beam can be expressed by

$$a = (W + x)\tan\alpha \approx (W + x)\sin\alpha \approx (W + x)\alpha, \tag{2.18}$$

where a is the distance between the spot B of the reflected light on the detector (position-sensitive) plane and the corresponding position A of the laser beam. In Eq. (2.18), the approximation of $\cos\alpha \approx 1$ and $\sin\alpha \approx \alpha$ can be made because of

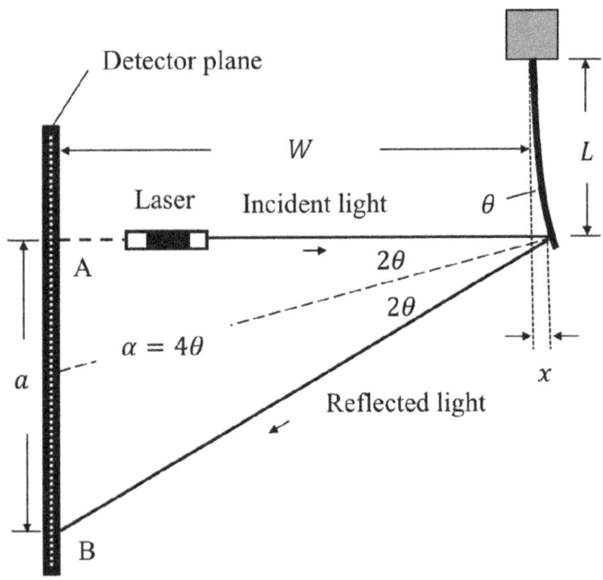

Fig. 2.10 Optical configuration for the cantilever bending setup in vacuum or air [1, 24]. In Fig. 2.10, a is the distance between the spot B of the reflected light on the detector (position-sensitive) plane and the corresponding position A of the laser beam. The distance x between the clamped end and reflection point of the cantilever is negligibly small as compared to the distance W between the clamped end of the cantilever and the detector plane. Modified from [24], Copyright 2010, with permission from AIP Publishing

$\alpha = 4\theta \approx 0$. The distance x between the clamped end and reflection point of the cantilever is negligibly small as compared to the distance W between the clamped end of the cantilever and the detector plane. Equation (2.18) leads to

$$\theta \approx \frac{a}{4W}. \tag{2.19}$$

The substitution of Eq. (2.19) into Eq. (2.17) provides the following relationship [1]:

$$\frac{1}{R} = \kappa \approx \frac{a}{2LW}. \tag{2.20}$$

Equation (2.20) means that the position of the reflected light spot on the detector plane moves in response to the change in bending of the cantilever (i.e., $\Delta\left(\frac{1}{R}\right)$ or $\Delta\kappa$). Therefore, the displacement Δa of the reflected light spot on the detector plane due to the change in bending of the cantilever is given by [1, 24]:

$$\Delta\left(\frac{1}{R}\right) = \Delta\kappa \approx \frac{\Delta a}{2LW}. \tag{2.21}$$

The optical detection of the cantilever bending has been used as a common technique for measurement of the surface stress change of a solid electrode in electrolyte solution [18, 25–32], which needs an electrochemical cell with an optical window. In an electrochemical system, a thin metal or semiconductor film coated on one side of a rectangular cantilever plate consisting of a glass or Si wafer, mica, etc., is mostly used as an electrode and the changes in surface stress can be measured from the change in the curvature (reciprocal of the curvature radius) of the cantilever as a function of the electrode potential or the surface charge density. Figure 2.11 shows the schematic representation and optical configuration of the cantilever bending setup in an electrochemical system [1, 33]. The problem in the electrochemical system is that the refraction of a laser light occurs at the optical window due to the difference in refractive index between air and solution.

As seen from Fig. 2.11, if the incident light is exactly normal to the optical window plane (or to the air/solution interface), no refraction of the incident light occurs at the optical window. Nevertheless, even if the normal incidence of the laser light, the direction of the light reflected by the surface of the cantilever is not normal to the optical window plane because of the bending of the cantilever. As a result, the refraction of the reflected light occurs at the optical window and the direction of the reflected light after passing the optical window shifts downward due to the low refractive index of air as compared to solution. The position of the spot of the reflected light on the position-sensitive detector also shifts downward ($B' \rightarrow B$). This means that the refraction of the reflected light at the optical window has to be taken into consideration for the exact determination of the curvature or curvature radius of the cantilever in the electrochemical system [1, 33]. In Fig. 2.11, the distances between the optical window and the cantilever and between the spots of the incident light

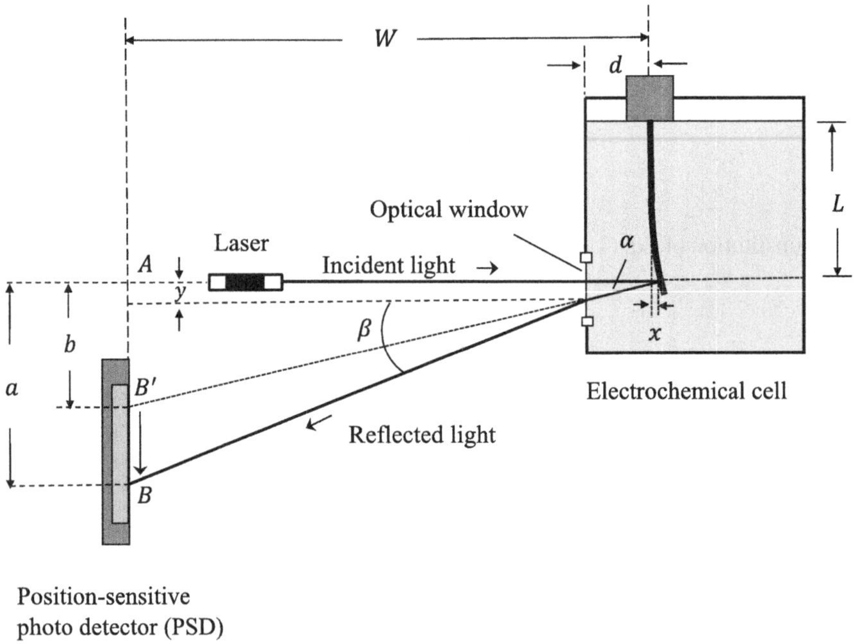

Fig. 2.11 Schematic representation and optical configuration of the cantilever bending setup in an electrochemical system [1, 33]. The distances between the optical window and the cantilever and between the spots of the incident light and the reflected light at the optical window are denoted by *d* and *y*, respectively. In addition, the distances between *A* and *B* and between *A* and *B′* are denoted by *a* and *b*, respectively. Modified from [33], Copyright 2000, with permission from Elsevier

and the reflected light at the optical window are denoted by d and y, respectively. In addition, the distances between A and B and between A and B' are denoted by a and b, respectively.

If the total deflection angle α is very small, the curvature radius of the cantilever can be obtained from Eq. (2.17):

$$R \approx \frac{2L}{\alpha}. \tag{2.22}$$

In the small-angle approximation, α under the conditions of $W \gg d \gg x$ and $b \gg y$, can be represented by

$$\alpha \approx \tan\alpha = \frac{b - y}{W - d} \approx \frac{b}{W}. \tag{2.23}$$

Furthermore, the angle β depicted in Fig. 2.11 can be also represented by

$$\beta \approx \tan \beta = \frac{a - y}{W - d} \approx \frac{a}{W}. \tag{2.24}$$

Since the position B on the SPD is really measured as a spot of the reflected light, it is necessary to know the relationship between α and β in order to obtain the curvature radius from Eq. (2.22).

The relationship between α and β is derived from the law of refraction (i.e., Snell's law) at the interface between two media of different refractive indices:

$$\frac{\sin \alpha}{\sin \beta} = \frac{n_a}{n_s}, \tag{2.25}$$

where n_a and n_s are the refractive indices of air and solution, respectively. For small deflections, Eq. (2.25) leads to

$$\frac{\sin \alpha}{\sin \beta} \approx \frac{\alpha}{\beta} \approx \frac{1}{n_{s,a}}, \tag{2.26}$$

where $n_{s,a}$ is the refractive index of the solution with respect to air, and it is practically equal to n_s because of $n_a = 1.0003$. Furthermore, α can be obtained from Eqs. (2.24) and (2.26):

$$\alpha \approx \frac{a}{n_{s,a} W}. \tag{2.27}$$

Thus, the curvature radius of the cantilever R is eventually given by [1, 33]:

$$R \approx \frac{2 n_{s,a} L W}{a}. \tag{2.28}$$

Equation (2.13) or Eq. (2.14) proves that the change in surface stress Δg is proportional to the change in reciprocal curvature radius $\Delta\left(\frac{1}{R}\right)$. If applying Eq. (2.28) to two different deflections of the cantilever, the following relationship is obtained:

$$\Delta\left(\frac{1}{R}\right) \approx \frac{\Delta a}{2 n_{s,a} L W}, \tag{2.29}$$

where Δa is the displacement of the reflected light spot on the PSD due to the change in bending of the cantilever electrode. The refractive indexes of aqueous solutions are in the range of $n_{s,a} = 1.33–1.48$. The neglect of the refraction of the reflected light at the optical window in the electrochemical system causes an error of about 25–30% for the determination of $\Delta\left(\frac{1}{R}\right)$ or Δg [1, 33].

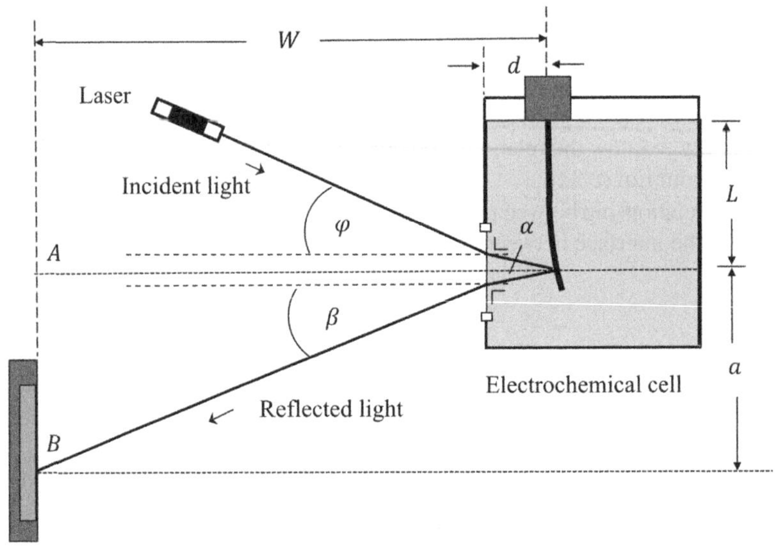

Fig. 2.12 Optical configuration of the cantilever bending experiments with non-normal incidence ($\varphi \neq 0$) in an electrochemical system [1, 34]. Modified from [34], Copyright 2005, with permission from Elsevier

In the case where the incident light is not normal to the optical window plane as shown in Fig. 2.12, the refraction of the incident light at the optical window in addition to the refraction of the reflected light has to be taken into consideration for the determination of the curvature radius [1, 34]. Although the calculation procedure is abbreviated here, the following relationship between $\Delta\left(\frac{1}{R}\right)$ and Δa has been derived by Rokob and Láng [34]:

$$\Delta\left(\frac{1}{R}\right) \approx \frac{\Delta a}{2n_{\mathrm{s,a}}LW}\left\{\frac{\left[1-(\sin\varphi)^2\right]^{3/2}}{\left[1-n_{\mathrm{s,a}}^{-2}(\sin\varphi)^2\right]^{1/2}}\right\} = \frac{\Delta a}{2n_{\mathrm{s},a}LW}\eta\left(\varphi, n_{\mathrm{s,a}}\right), \tag{2.30}$$

where φ is the incidence angle of the laser light at the optical window plane. The effect of the incident angle is represented by the term of $\eta\left(\varphi, n_{\mathrm{s,a}}\right)$ in Eq. (2.30) which decreases monotonously with increasing φ. For example, the value of $\eta\left(\varphi, n_{\mathrm{s,a}}\right)$ is 0.86 at $\varphi = 20°$ for $n_{\mathrm{s,a}} = 1.333$ (pure water at 20 °C) [1, 34]. In the case of normal incidence (i.e., $\varphi = 0$), the value of $\eta\left(\varphi, n_{\mathrm{s,a}}\right)$ becomes unity, and thus, Eq. (2.30) is returned to Eq. (2.29). Equation (2.30) is useful to estimate an error in the value of $\Delta\left(\frac{1}{R}\right)$ due to an optical misalignment resulting in non-normal incidence of the laser light.

When φ is within 2°, the error is less than 0.15%. On the other hand, the error attains 4% in the case of $\varphi = 10°$. Furthermore, for the accurate calculation of $\Delta\left(\frac{1}{R}\right)$, the dependence of $n_{s,a}$ on light wavelength and temperature have to be taken into consideration. In the derivation of Eq. (2.28) or Eq. (2.29), the thickness of the optical window (made from quartz or fused silica) d_w is neglected. The light reflected from the mirror surface of the cantilever in solution is refracted at the optical window side which is faced to the solution, and then the refracted light travels inside the optical window. When the light is passing from the optical window to air, the refraction takes place again at the optical window/air interface. As a result, the refraction in twice at the optical window produces the lateral shift (or lateral displacement) l_s of the reflected light on the detector plane (PSD). The following relationship of the relative error $\frac{\Delta l_s}{\Delta a}$ for the neglect of the thickness of optical window d_w was derived from the calculation of the optical configuration for the lateral shift [1]:

$$\frac{\Delta l_s}{\Delta a} \approx \frac{d_w}{W}\left(n_{w,a} - 1\right), \tag{2.31}$$

where Δl_s is the change in lateral displacement of the reflected light on the detector plane due to the change in bending of the cantilever electrode and $n_{w,a}$ is the refractive index of the optical window with respect to air. The symbols of Δa and W in Eq. (2.31) have the same meanings as those used in Eq. (2.29). In a typical experiment of the cantilever bending, W is in the range of 1 m and d_w is in the range of 1 mm. If a fused quartz ($n_{w,a} = 1.46$) is used as an optical window, $\frac{\Delta l_s}{\Delta a} \approx 4.6 \times 10^{-4}$ is estimated. In this case, the error caused by the lateral shift at the optical window can be safely neglected [1].

Dynamic Stress Analysis (DSA) of Cantilever Bending

The changes in surface stress of the electrode induced by potential- or surface charge-modulation can be measured from the changes in the curvature of the cantilever as a function of frequency [35, 36], which is somewhat similar to the piezoelectric detection of differential surface (see Sect. 2.2 of this chapter). This method is named "dynamic stress analysis (DSA)," which is achieved by the combination of electrochemical impedance spectroscopy (EIS) and mechanical impedance spectroscopy of the cantilever electrode. In DAS experiment as well as EIS, a small sinusoidal voltage $E_o \exp(j\omega t)$ is superimposed on a DC potential E_{dc}. The potential E applied to the cantilever electrode is expressed by

$$E = E_{dc} + E_o \exp(j\omega t), \tag{2.32}$$

where E_o is the potential amplitude, ω is the angular frequency, and $j = \sqrt{-1}$. The corresponding current density response is formulated by

$$i = i_{dc} + i_o \exp\left[j(\omega t + \psi_e)\right], \tag{2.33}$$

where i_o is the current density amplitude and ψ_e is the phase angle between current and potential. Similarly, the corresponding surface stress response, i.e., the mechanical response, of the cantilever electrode is given by

$$g = g_{dc} + g_o \exp\left[j(\omega t + \psi_s)\right], \tag{2.34}$$

where g_o is the surface stress amplitude and ψ_s is the phase angle between surface stress g and potential E.

Figure 2.13 demonstrates the schematic data predicted from dynamic stress analysis (DSA): (a) the current density i and (b) the surface stress g responses for a sinusoidal potential modulation of a noble metal electrode at a certain DC potential E_{dc} in the electric double-layer region with values of $\psi_e \approx -90°$ and of $\psi_s \approx -180°$ [35]. The AC component q_{ac} of the surface charge density can be obtained by integrating the second term in the right-hand side of Eq. (2.33) with respect to time. The AC component g_{ac} of surface stress divided by the AC component q_{ac} of surface charge density; i.e., $\frac{g_{ac}}{q_{ac}}$ is equivalent to $\frac{\partial g}{\partial q}$, and the phase angle of surface stress

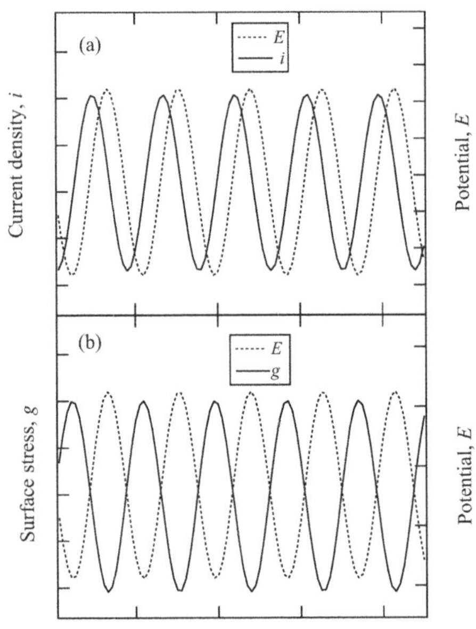

Fig. 2.13 Schematic data predicted from dynamic stress analysis (DSA): **a** the current density i and **b** the surface stress g responses for a sinusoidal potential modulation of a noble metal electrode at a certain DC potential E_{dc} in the electric double-layer region with values of $\psi_e \approx -90°$ and of $\psi_s \approx -180°$. ψ_e and ψ_s are the phase angles between i and E, and between g and E, respectively

relative to the surface charge density represents the sign of $\frac{\partial g}{\partial q}$. Therefore, the value of $\frac{\partial g}{\partial q}$ obtained as a function of potential by DSA [35, 36] is comparable with that of $\frac{\partial g}{\partial q}$ obtained from the q-estance method [2]. The DSA method is useful to evaluate the electro-chemo-mechanical properties of solid electrode surfaces. The details and typical application results of DSA are described and discussed in the Sect. 3.4.2 of Chap. 3.

2.4 Elastic Deformation of Metal Electrode Associated with Surface Stress

An extensometer is capable of measuring the changes in surface stress from the relative changes in length of a very thin metal ribbon or wire (as a working electrode) as a function of electrode potential in electrolyte solution. The principle and details of the extensometer method have been described elsewhere [1, 13, 37]. In addition, the stress effect on open circuit potential [38, 39] has been investigated by applying a force to a very thin metal wire in electrolyte solution which is quite similar to the extensometer method.

The experimental setup is schematically shown in Fig. 2.14 [38]. An Ag wire (0.22 mm in diameter) is made taut vertically in 0.25 M KNO_3 solutions containing 0.01–0.1 M $AgNO_3$. The bottom of the vessel, which contains the wire, is closed with a silica plug. After the wire was led through the plug by using a hypodermic needle,

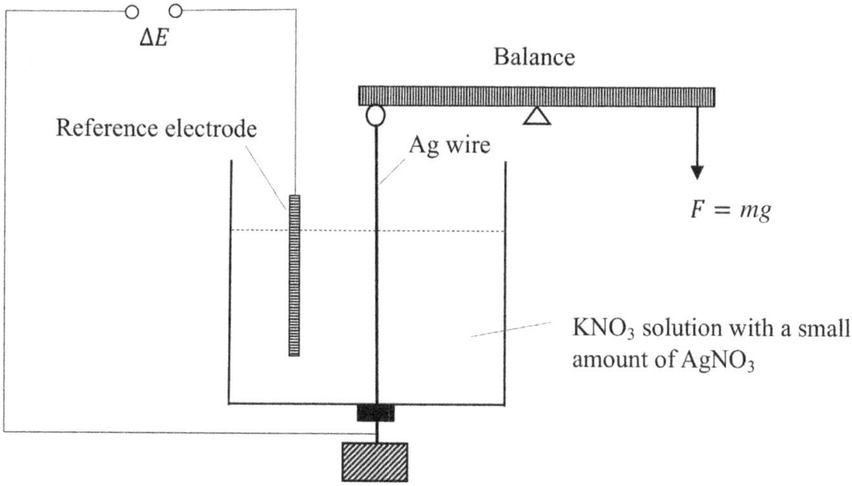

Fig. 2.14 Schematic view of the experimental setup for measuring the stress effect on open circuit potential of an Ag wire electrode in KNO_3 solution containing a small amount of $AgNO_3$ [38]. Reproduced from [38] with permission from the PCCP Owner Societies

the remaining pinhole is sealed with silicone grease. The upper end of the wire is fixed to one arm of a balance, the other arm of which is loaded with standard weights. The total length of the wire is 50 mm, the portion in contact with the electrolyte solution being 30 mm. The strain ε is equal to the relative change in wire length $\frac{dl}{l}$, while the stress σ is given by a force per cross-sectional area of the wire, i.e., $\sigma = \frac{F}{A} = \frac{mg}{A}$ (F: force, m: mass, g: acceleration of gravity, and A: cross-sectional area). The electrochemical cell is housed in a Faraday cage. The potential difference between the Ag wire and the reference electrode is measured with a voltmeter with a high input resistance of $10^9\ \Omega$. The applied force is sufficiently low to keep proportionality between stress and strain (within a limit of elastic deformation). The change of open circuit potential ΔE_{ocp} is determined as a function of σ under quasi-static strain by load on-load off cycles. The value of $\frac{\mathrm{d}E_{\mathrm{ocp}}}{\mathrm{d}\sigma} = 2.6 \times 10^{-11}$ V Pa^{-1} was obtained from the linear relation between ΔE_{ocp} and σ, which was converted to $\frac{\mathrm{d}E_{\mathrm{ocp}}}{\mathrm{d}\varepsilon} = 1.04$ V by using the isothermal elastic compliance $s = \frac{\mathrm{d}\varepsilon}{\mathrm{d}\sigma} = 2.5 \times 10^{-11}$ Pa^{-1} at room temperature for the Ag wire [38].

Dynamic Electro-Chemo-Mechanical Analysis (DECMA)

As an alternative method, potential changes during cyclic elastic deformation of a metal thin film on a polymer substrate can be directly measured by using a lock-in technique [40–42]. Figure 2.15 shows (a) the scheme of the experimental setup for the potential variation during cyclic elastic deformation and (b) the top view of the working (Au thin film) electrode used for the experiment [41]. A (111)-textured Au thin-film (20 nm) electrode was prepared by DC magnetron sputtering onto the top of a thin titanium adhesion layer (2 nm) on a thick polyimide (Kapton) sheet (125 μm) with Poisson's ratio of $\nu = 0.35$. The electrode geometry was achieved by sputtering through a shadow mask. The polyimide substrate with the Au film electrode is attached to a fixed grip (right) and a mobile grip (left) which can be cyclic displaced by a computer-controlled piezoactuator equipped with a calibrated displacement sensor. The response of the actuator limits the frequency range of cyclic strain to $\omega \leq 100$ Hz. The Au film facing down is wetted from below by a meniscus of solution.

The dashed line in Fig. 2.15b represents the wetting boundary so that the wetted region is larger than the circular electrode section, assuring constant wetted Lagrangian area (constant number of surface metal atoms in contact with solution) throughout the strain cycles. When the axial strain of $\frac{dl}{l}$ (l: gauge length) is imposed on the substrate, the area strain of the substrate is given by $\mathrm{d}\varepsilon = \frac{(1-\nu)dl}{l}$. This strain is precisely transferred to the Au electrode, which has been confirmed by in situ diffraction under load [43]. The reference electrode is separated from the main body of the electrochemical cell by a Luggin capillary. Au wire as the counter electrode is set in a compartment separated from the main reservoir by a channel. The entire equipment is housed in a stainless steel chamber with fittings of high-vacuum grade. Before experiments, the chamber is flushed repeatedly with a high-purity argon gas and then sealed in argon gas at atmospheric pressure.

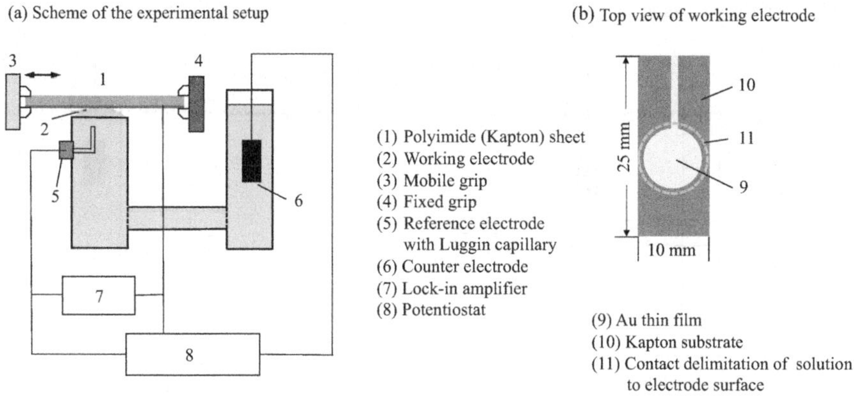

Fig. 2.15 **a** Scheme of the experimental setup for the potential variation during cyclic elastic deformation and **b** the top view of the working (Au thin film) electrode used for the experiment [40]. Reproduced from [40] with permission from the PCCP Owner Societies

The potential of the Au electrode in 0.01 M $HClO_4$ solution is measured by a potentiostat, and the variation of the open circuit potential E_{ocp} during cyclic elastic strain is recorded by a lock-in amplifier which adopts the piezodrive signal as a trigger. A sinusoidal displacement of the mobile grip $dl(t) = \varepsilon_o l \sin(\omega t)$ with peak-to-peak amplitude $2l\varepsilon_o$ is applied up to $dl = 20$ μm at a gauge length of $l = 24$ mm in the frequency region between $\omega = 0.3$ and 100 Hz. The elastic strain amplitude is less than 8.3×10^{-4}. The amplitude of $\frac{\partial E}{\partial \varepsilon}$ (with a negative value) obtained by the above experiment increases with increasing frequency up to 1 Hz and then attains to a saturated value of about 1.8 V beyond 30 Hz. The low negative value of $\frac{\partial E}{\partial \varepsilon}$ below 30 Hz was ascribed to the discharging of the electric double-layer capacity due to Faraday current [40, 41]. In addition to the measurement of the potential–strain response under open circuit conditions (Fig. 2.15), the measurements of the current–strain response and of the potential–strain response in a potentiostatic mode can be achieved by inserting a shunt resistance R_S and a delay resistance R_D, respectively, as shown in Fig. 2.16 [42].

In the current–strain response in a potentiostatic mode (Fig. 2.16a), a shunt resistance of $R_S = 46\ \Omega$ is inserted between the potentiostat and counter electrode, and the current magnitude is determined from the magnitude of the potential drop across R_S. The cyclic voltammetry of the Au electrode in 0.01 M $HClO_4$ solution is performed at a potential scan rate of 1–10 mVs^{-1}, while the electrode is subjected to cyclic strain under a frequency of 20 Hz and an amplitude of $\varepsilon_o = 2 \times 10^{-4}$. This strain cycle is sufficiently slow for the potentiostat to compensate the strain-induced potential change. In contrast, the strain-induced current variation is well resolved, in spite of the small strain amplitude. Provided that the electrochemical impedance of the electrode Z_e is known, the value of $\frac{\partial E}{\partial \varepsilon}$ can be determined as a function of potential from the current density amplitude i_o[42].

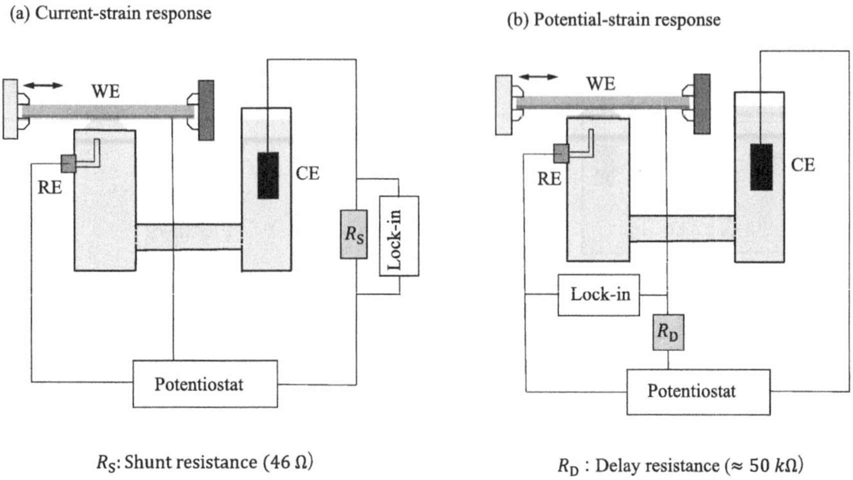

Fig. 2.16 Scheme of the experimental setup for **a** the current–strain response and for **b** the potential–strain response in a potentiostatic mode during cyclic elastic deformation [42]. Reproduced from [42] with permission from the PCCP Owner Societies

In the potential–strain response in a potentiostatic mode (Fig. 2.16b), a delay resistance of $R_D \approx 50\text{k}\Omega$ is connected in series with the working electrode, which leads to an increase in the time constant for potential control at the working electrode, preventing the potentiostat from compensating the potential change due to the cyclic strain. On the other hand, the current oscillation due to the cyclic strain becomes negligible. As a result, the value of $\frac{\partial E}{\partial \varepsilon}$ can be directly determined from the strain-induced potential change measured by a lock-in amplifier connected between the working electrode and the reference electrode. Furthermore, in a galvanic mode in which the current is controlled rather than the potential, the measurement of the potential–strain response can be achieved without inserting a delay resistance. A galvanic cycle is performed between the current limits of -1 and 1 μA at a current scan rate of 10 nA s^{-1} under the applied cyclic strain. This mode is directly linked to a potential change with strain at constant surface charge density since q and ε are controlled and E is measured. The values of $\frac{\partial E}{\partial \varepsilon}$ obtained by three different modes in Fig. 2.16 coincided each other within a small error in the entire potential range [42]. The above analysis with three different modes is named "dynamic electro-chemo-mechanical analysis (DECMA)" [42]. As represented by Eq. (1.124) in Sect. 1.8 of Chap. 1, $\left(\frac{\partial E}{\partial \varepsilon}\right)_q$ is equivalent to $\left(\frac{\partial g}{\partial q}\right)_\varepsilon$. The details and typical application results of DECMA are described in Sect. 3.4.3 of Chap. 3. Moreover, we discuss the equality between the experimental values of $\frac{\partial E}{\partial \varepsilon}$ and $\frac{\partial g}{\partial q}$, obtained independently by DECMA, DSA and other methods in Sect. 3.4 of Chap. 3.

2.5 Dilatometric Detection of Strain Change for Nano-Porous Metal Electrode

In situ dilatometry has been developed for measurement of the changes in strain of a nano-porous metal electrode as a function of potential in electrolyte solution [44, 45]. Furthermore, the changes in strain can be converted to the change in surface stress if the bulk modulus and specific surface area of a nano-porous metal are known [46, 47]. The nano-porous Pt electrode [44, 46, 47] was prepared by consolidating commercial Pt black having a grain size of 6 nm, while the nano-porous Au electrode [45, 48, 49] was prepared by dealloying a master alloy $Ag_{75}Au_{25}$ (atomic%) in 1 M $HClO_4$ solution under potentiostatic control. The nano-porous Au–Pt alloys can be also prepared by dealloying homogeneous master alloys $(Au_{1-x}Pt_x)_{25}Ag_{75}$ in 1 M $HClO_4$ solution [50].

Scanning electron micrograph (SEM) of the fracture surface of the nano-porous Pt sample showed the bi-continuous structure, consisting of an interconnected network of nanometer-sized Pt crystallites and an interconnected pore space [44, 46, 47]. Similarly, the SEM of the nano-porous Au sample showed the bi-continuous (sponge-like open-cell) structure with a ligament size of about 20 nm while maintaining the original shape of the alloy sample [48]. Figure 2.17 represents a schematic illustration of the experimental setup for in situ dilatometry of a nano-porous metal electrode [48]. The nano-porous Pt electrode (e.g., 1.8 mm in length and 1.5 mm in diameter) is set in the sample space of a commercial dilatometer (Netzsch® 402C) encapsulated in a miniaturized electrochemical cell [47]. The sample compartment, which contains the

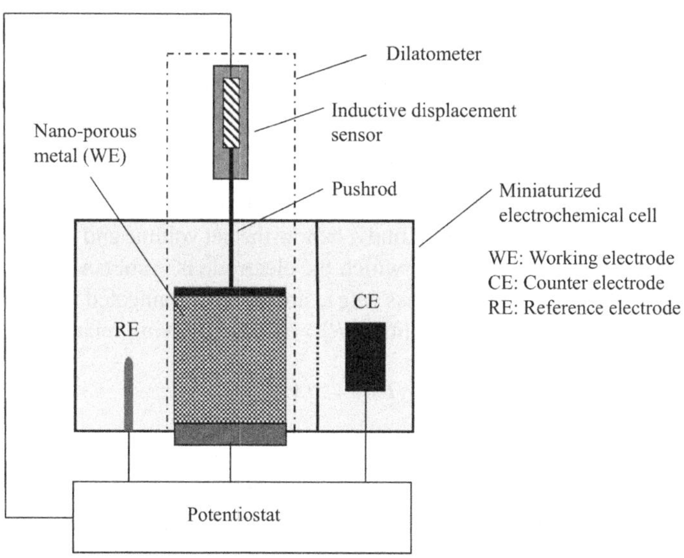

Fig. 2.17 Schematic illustration of the experimental setup for in situ dilatometry of a nano-porous metal electrode [48]. Modified from [48], Copyright 2008, with permission from Elsevier

commercial reference electrode (Ag/AgCl/3 M KCl), is separated from the counter electrode compartment by a glass frit [48]. The reference electrode is in a Teflon tube with a porous membrane (ceramic/conducting polymer) at the junction with low electrolyte leakage. The counter electrode is a porous Pt which is the same material as the samples. The changes in sample length are transmitted to an inductive displacement sensor via a pushrod (made from silica) loaded by a contact pressure of 20 cN [47]. The sample space of the dilatometer is connected to a bath thermostat to maintain a steady-state temperature of 283 ± 0.1 K. For synchronization between dilatometric detection and electrochemical measurement, the output signal of the dilatometer is supplied to the external input channel of the potentiostat and then is saved together with current and net charge in a same data file [47].

The mechanical equilibrium between the bulk and surface of a solid is brought by balancing the forces acting at the surface with the stress in the bulk. The equilibrium condition for the case of a fluid droplet is the Young–Laplace equation [51, 52], which is expressed by

$$\Delta P = P - P_{\mathrm{o}} = \frac{2\gamma}{R}, \tag{2.35}$$

where P_{o} is the external pressure, and P, γ, and R are the internal pressure, surface tension, and curvature radius of the fluid droplet, respectively. The pressure difference ΔP (i.e., $\frac{2\gamma}{R}$) in Eq. (2.35) is the thermodynamic driving force for formation of the fluid droplet. The situation in a solid is quite different from that in a fluid. The equilibrium condition at a curved surface in a solid is linked to the surface stress g in place of surface tension γ, since the solid is subjected to shear. Weissmüller and Cahn [53] derived a generalized capillary equation for a solid, which is formulated by

$$3V\langle P - P_o\rangle_V = 2A\langle g\rangle_A, \tag{2.36}$$

where the angular brackets designate averages over the volume of the bulk solid and over the entire surface area, while V and A denote the net volume and surface area, and P_{o} is the pressure in the solution in which the electrode is immersed. Consequently, the changes in the mean surface stress $\langle \Delta g\rangle_A$ are directly connected with the changes in the mean pressure in the bulk solid $\langle \Delta P\rangle_V$ by the following relationship:

$$3V\langle \Delta P\rangle_V = 2A\langle \Delta g\rangle_A. \tag{2.37}$$

It has been confirmed by an atomistic simulation study [54] that Eq. (2.36) can be applied to microstructures of different geometry which include (1) spherical particles with convex surfaces, (2) solids containing an array of spherical voids with concave surfaces, and (3) bi-continuous network structures with interpenetrating solid phase and pore space. If Eq. (2.37) is applied to a nano-porous metal electrode, the changes in mean surfaces stress $\langle \Delta g \rangle_A$ can be determined from the relative changes in sample length measured by dilatometry. The volumetric mean strain is given by

$$\frac{\Delta V}{V} = -\frac{\langle \Delta P \rangle_V}{Y_\mathrm{b}}, \tag{2.38}$$

where Y_b is the bulk modulus of the metal. For small strain, the following relationship between the relative changes in volume and sample length holds:

$$\frac{\Delta V}{V} = \frac{3\Delta l}{l_\mathrm{o}}, \tag{2.39}$$

where $\frac{\Delta l}{l_\mathrm{o}}$ corresponds to the macroscopic strain of the sample. The specific surface area per volume or mass can be measured by Brunauer–Emmett–Teller (BET) method [55]. The substitution of Eqs. (2.38) and (2.39) into Eq. (2.37) leads to the relationship between $\langle \Delta g \rangle_A$ and $\frac{\Delta l}{l_\mathrm{o}}$ [46, 47]:

$$\langle \Delta g \rangle_A = -\frac{9Y_\mathrm{b}}{2\alpha}\frac{\Delta l}{l_\mathrm{o}} = -\frac{9Y_\mathrm{b}}{2\alpha_\mathrm{m}\rho}\frac{\Delta l}{l_\mathrm{o}}, \tag{2.40}$$

where α is the specific surface area per volume $\left(\mathrm{AV}^{-1}\right)$, α_m the specific surface area per mass ($\mathrm{m}^2\ \mathrm{g}^{-1}$), and ρ is the mass density ($\mathrm{g\ m}^{-3}$). We discuss the typical application results of the in situ dilatometry in Sect. 3.4.4 of Chap. 3.

References

1. Láng GG, Barbero CA (2012) Laser techniques for the study of electrode processes. chaps. 4 and 5. Springer, Berlin
2. Gokhshtein AY (1976) Surface tension of solids and adsorption. Nauka, Moscow
3. Gokhshtein AY (1970) Electrochim Acta 15:219–223
4. Gokhshtein AY (1975) Russ Chem Rev 44:921–932
5. Valincius G (1998) Langmiur 14:6307–6319
6. Valincius G (1999) J Electroanal Chem 478:40–49
7. Malpus RE, Fredlein RA, Bard AJ (1979) J Electroanal Chem 98:171–180
8. Handley LJ, Bard AJ (1980) J Electrochem Soc 127:338–343
9. Seo M, Makino T, Sato N (1986) J Electrochem Soc 133:1138–1142
10. Seo M, Jiang XC, Sato N (1987) J Electrochem Soc 134:3094–3098
11. Dickinson KM, Hansen KE, Fredlein RA (1992) Electrochim Acta 37:139–141
12. Bode-Jr DD, Andeson TN, Eyring H (1967) J Phys Chem 71:792–797
13. Lin K-F, Beck TR (1976) J Electrochem Soc 123:1145–1151
14. Jiang XC, Seo M, Sato N (1991) J Electrochem Soc 138:137–140
15. Seo M, Aomi M (1992) J Electrochem Soc 139:1087–1090

16. Seo M, Aomi M (1993) J Electroanal Chem 347:185–194
17. Seo M, Ueno K (1996) J Electrochem Soc 143:899–904
18. Ueno K, Seo M (1999) J Electrochem Soc 146:1496–1499
19. Seo M, Serizawa Y (2003) J Electrochem Soc 150:E472–E476
20. Stoney GG (1909) Proc Roy Soc London A 32:172–175
21. Ibach H (1997) Surf Sci Rep 29:193–263
22. Ibach H (2006) Physics of surfaces and interfaces. Springer, Berlin
23. Dahmen K, Lehwald S, Ibach H (2000) Surf Sci 446:161–173
24. Láng GG (2010) J Appl Phys 107:116104-1-3
25. Fredlein RA, Damjanovic A, Bockris JO'M (1971) Surf Sci 25:264–361
26. Sahu SN, Scarminio J, Decker F (1990) J Electrochem Soc 137:1150–1154
27. Láng GG, Ueno K, Ujvári M, Seo M (2000) J Phys Chem B 104:2785–2789
28. Seo M, Yamazaki M (2004) J Electrochem Soc 15:E276–E281
29. Láng GG, Seo M, Heusler KE (2005) J Solid State Electrochem 9:347–353
30. Stafford GR, Bertocci U (2006) J Phys Chem B 110:15493–15498
31. Smetanin M, Viswanath RN, Kramer D, Beckmann D, Koch T, Kibler LA, Kolb DM, Weissmüller J (2008) Langmuir 24:8561–8567
32. Van Overmeere Q, Vanhumbeek J-F, Proost J (2010) Rev Sci Instr 81:045106-1-10
33. Láng GG, Seo M (2000) J Electroanal Chem 490:98–101
34. Rokob TA, Láng GG (2005) Electrochim Acta 51:93–97
35. Lafouresse MC, Bertocci U, Beauchamp CR, Stafford GR (2012) J Electrochem Soc 159:H816–H822
36. Lafouresse MC, Bertocci U, Stafford GR (2013) J Electrochem Soc 160:H636–H643
37. Beck TR (1969) J Phys Chem 73:466–468
38. Horváth Á, Schiller R (2001) Phys Chem Chem Phys 3:2662–2667
39. Horváth Á, Nagy G, Schiller R (2010) Phys Chem Chem Phys 12:7290–7290
40. Smetanin M, Kramer D, Mohanan S, Herr U, Weissmüller J (2009) Phys Chem Chem Phys 11:9008–9012
41. Smetanin M, Deng O, Kramer D, Mohanan S, Herr U, Weissmüller J (2010) Phys Chem Chem Phys 12:7291–7292
42. Smetanin M, Deng Q, Weissmüller J (2011) Phys Chem Chem Phys 13:17313–17322
43. Özkaya B, Saranu SR, Mohanan S, Herr U (2008) Phys Stat Sol A 205:1876–1879
44. Weissmüller J, Viswanath RN, Kramer D, Zimmer P, Würschum R, Gleiter H (2003) Science 300:312–315
45. Kramer D, Viswanath RN, Weissmüller J (2004) Nano Lett 4:793–796
46. Viswanath RN, Kramer D, Weissmüller J (2005) Langmuir 21:4604–4609
47. Viswanath RN, Kramer D, Weissmüller J (2008) Electrochim Acta 53:2757–2767
48. Jin H-J, Parida S, Kramer D, Weissmüller J (2008) Surf Sci 602:3588–3594
49. Biener J, Wittstock A, Zepeda-Ruiz LA, Biener MM, Zielasek V, Kramer D, Viswanath RN, Weissmüller J, Bäumer M, Hamza AV (2009) Nat Mater 8:47–51
50. Jin H-J, Wang X-L, Parida S, Wang K, Seo M, Weissmüller J (2010) Nano Lett 10:187–194
51. Young T (1805) Philos Trans R Soc London 95:65–87
52. Laplace PS (1806) Celestial mechanics: translated in 1966 by Chelsea Pub Co., New York
53. Weissmüller J, Cahn JW (1997) Acta Mater 45:1899–1906
54. Weissmüller J, Duan H-L, Farkas D (2010) Acta Mater 58:1–13
55. Brunauer S, Emmett PH, Teller E (1938) J Am Chem Soc 60:309–319

Chapter 3
Potential- or Adsorbate-Induced Changes in Surface Stress of Solid Metal Electrode, and Surface Stress versus Surface Charge Density or Potential versus Surface Elastic Strain

Abstract Potential- or adsorbate-induced changes in surface stress associated with surface reconstruction or adsorption of electrolyte anions on single-crystal Au electrodes are explained and discussed as typical examples of the electro-chemo-mechanical properties. The surface stress versus potential curves measured for single-crystal Au electrodes subjected to surface reconstruction or adsorption of electrolyte anions are compared with the corresponding electrocapillary (surface tension vs. potential) curves, and it is experimentally confirmed that surface stress differs from surface tension for a solid electrode. Two important parameters for characterizing solid electrode surfaces are surface stress–surface charge coefficient and potential–surface elastic strain coefficient which are theoretically equivalent to each other. The above two coefficients measured separately by several methods for a (111)-textured Au (111) thin-film electrode in perchloric acid solution are compared, and the equality of both coefficients is confirmed experimentally. Furthermore, the sign-reversals of the above coefficients observed for nano-porous or (111)-textured Pt electrodes in acid solutions are explained and the causes of the sign-reversal are discussed on the basis of the changes in electronic structure of the electrode surfaces.

Keywords Surface stress · Surface reconstruction · Adsorption · Surface stress–surface charge density coefficient · Potential–surface elastic strain coefficient

3.1 Introduction

Potential- or adsorbate-induced changes in surface stress associated with surface reconstruction or adsorption of electrolyte species on a solid electrode are typical examples of the electro-chemo-mechanical properties. The changes in surface stress of single-crystal Au electrodes for surface reconstruction or adsorption of electrolyte anions have been measured as function of applied potential or surface charge density by a cantilever bending method. On the other hand, the concomitant changes in surface tension of single-crystal Au electrodes can be estimated by using the Lippmann equation from the measurement of differential capacity or surface charge density. In the present chapter, we compare the surface stress versus potential curves

M. Seo, *Electro-Chemo-Mechanical Properties of Solid Electrode Surfaces*,
https://doi.org/10.1007/978-981-15-7277-7_3

measured for surface reconstruction or adsorption of electrolyte species on single-crystal Au electrodes with the corresponding electrocapillary (surface tension vs. potential) curves to make clear the difference between surface stress and surface tension.

The changes in surface stress depend not only on potential but also on surface charge density. The dependence of surface stress g on surface charge density q, i.e., $\zeta_{g,q} = \frac{\partial g}{\partial q}$, is named "surface stress–surface charge density coefficient," which is characteristic for solid electrode surfaces. The changes in surface elastic strain of a solid electrode induce the changes in electrode potential. The dependence of potential E on surface elastic strain ε, i.e., $\zeta_{E,\varepsilon} = \frac{\partial E}{\partial \varepsilon}$, is named "potential–surface elastic strain coefficient," which is also characteristics for solid electrode surfaces. It has been theoretically derived that both coefficients are equivalent to each other (see Eq. (1.124) in Sect. 1.8 of Chap. 1). The above two characteristic parameters can be measured separately by several methods such as DSA and DECMA (see Sects. 2.3 and 2.4 of Chap. 2). We compare the values of $\zeta_{g,q}$ and $\zeta_{E,\varepsilon}$ measured separately for a (111)-textured Au (111) film electrode in perchloric acid solution to confirm that $\zeta_{g,q}$ is equivalent to $\zeta_{E,\varepsilon}$. The sign-reversal of $\zeta_{g,q}$ or $\zeta_{E,\varepsilon}$ has been observed for nano-porous or (111)-textured Pt electrodes in acid solutions, depending on the potential regions of hydrogen adsorption, electric double layer (or oxygen adsorption), and oxide formation. We discuss the origin of the sign-reversal on the basis of the changes in electronic structure of the electrode surfaces.

3.2 Surface Reconstruction

When a clean solid surface is created by cleavage in ultra-vacuum, the surface atoms have an excess energy (i.e., surface energy) as compared to that of interior atoms in solid since the number of the nearest-neighbor atoms decreases on the surface to bring the increase in electronic charge that does not participate in bonding. As a result, surface relaxation in which the bond length normal to the surface varies slightly for atoms in the surface region takes place on the clean surface to minimize the surface energy. Particularly, the clean surfaces of the face-centered cubic (fcc) transition metals such as Au, Pt, and Pd are often subjected to surface reconstruction in which surface atoms undergo a lateral displacement to form a two-dimensional superlattice different from the interior lattice structure [1–3].

For example, the (100) surface of fcc transition metal reconstructs into a hexagonal close-packed (hcp) form due to a heat treatment (see Fig. 3.1a) [1–3]. The reconstructed (100) surface is slightly buckled because of the significant structural misfit between the hcp top layer and the underlying (100) plane, and it is named "hex" structure [1, 2]. The atomic density of the reconstructed hex structure is higher by 20–25% than the unreconstructed (100) surface. On the other hand, the (110) surface reconstructs into a (1×2) missing-row structure, where every second row in [001] direction is missing (see Fig. 3.1b) [1, 2, 4, 5]. The surface reconstruction of the (100) and (110) surfaces accompanies the increase in atomic density of surface

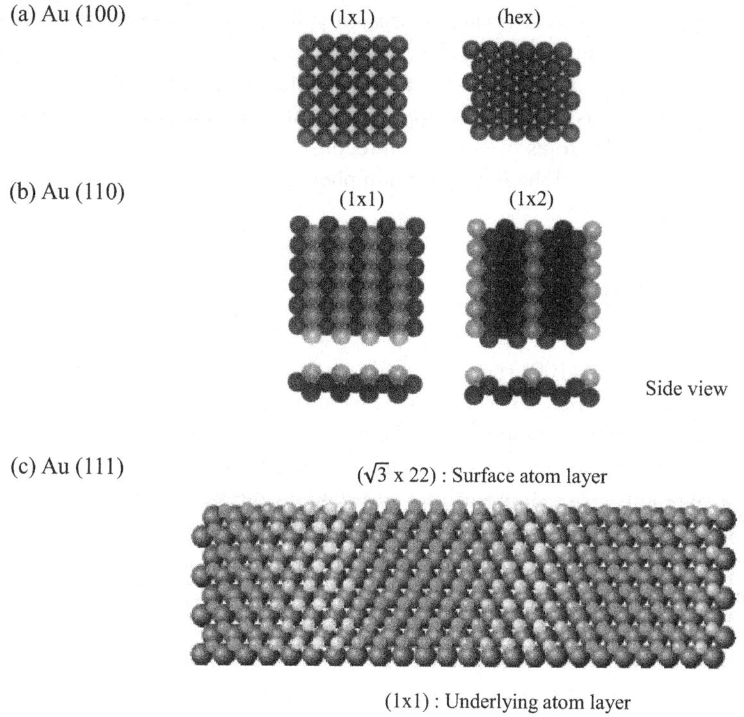

Fig. 3.1 Structures of unreconstructed and reconstructed gold single-crystal surfaces [2]. **a** Au (100): $(1 \times 1) \leftrightarrow (\text{hex})$, **b** Au (110): $(1 \times 1) \leftrightarrow (1 \times 2)$, and **c** Au (111): $(1 \times 1) \leftrightarrow (\sqrt{3} \times 22)$. Reprinted from [2], Copyright 1996, with permission from Elsevier

layer which reduces the surface energy due to the decrease in electronic charge that does not participate in bonding.

In the case of the (111) surface, the (111) surface of Pt does not spontaneously reconstruct at temperature lower than 1330 K [6], while the Au (111) surface reconstructs into a $\left(\sqrt{3} \times 22\right)$ structure in which every 23rd surface atom is in register with every 22nd atom of the underlying lattice, involving uniaxial compression by about 4% in the surface layer along the [110] direction [1, 2]. The net change in surface structure due to reconstruction is relatively small as compared to the (100) and (110) surfaces since the unreconstructed (111) surface has the already densely packed structure. As shown in Fig. 3.1c, the $\left(\sqrt{3} \times 22\right)$ surface has a complicated structure in which the position of surface atoms with respect to the second layer changes periodically between two different domains of fcc sites and hcp sites through a domain wall of bridge sites [7].

In interfacial electrochemistry, it is known that the transition between reconstructed and unreconstructed surfaces for single-crystal electrodes such as Au and Pt takes place, depending on applied potential and adsorption of electrolyte anions [1,

2]. Surface stress g as well as surface tension γ may play vital roles in reconstruction of the electrode surfaces. The changes in surface stress for the reconstructed and unreconstructed surfaces for Au (100) and (111) electrodes in 0.1 M $HClO_4$ solution have been measured as a function of applied potential by a cantilever bending method [8, 9]. We discuss the roles of g and γ in reconstruction for the Au (100) and (111) electrodes to understand the reconstruction phenomena.

3.2.1 Au (100) Surface

The reconstructed Au (100)-(hex) electrode surface is transformed (i.e., rifted) to the unreconstructed Au (100)-(1 × 1) surface by anodic polarization at potentials more positive than a critical potential [1, 2]. The critical potential for rifting can be determined from the potential at a spike of anodic current during anodic potential scan. The critical potential of rifting shifts toward negative direction in accordance with the order of the adsorption strength of electrolyte anions, $Cl^- > HSO_4^- > ClO_4^-$ [10]. The reconstructed surface, however, is restored by cathodic polarization at potentials more negative than the critical potential after the lifting to the unreconstructed surface. The critical potential for restoration to the reconstructed surface does not depend sensitively on anion species [1, 2]. The potential of zero charge E_{pzc} of a solid electrode can be determined from the location of a differential capacity minimum in the potential region of electric double layer.

Figure 3.2 shows the differential capacity versus potential (c vs. E) curves in the potential region of electric double layer between −0.4 and 0.6 V (SCE) for the Au

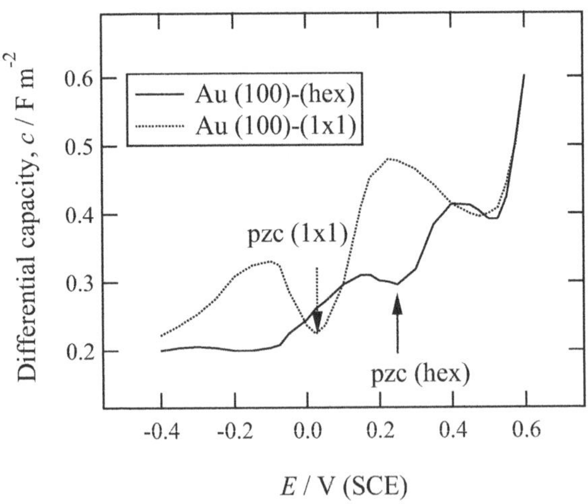

Fig. 3.2 Differential capacity versus potential (c vs. E) curves in the potential region of electric double layer between −0.4 and 0.6 V (SCE) for the Au (100)-(hex) and -(1 × 1) surfaces in 0.01 M $HClO_4$ solution [11]. The vertical arrow shows the location of potential of zero charge (pzc). Reprinted from [11], Copyright 1996, with permission from Elsevier

(100)-(hex) and -(1 × 1) surfaces in 0.01 M $HClO_4$ solution [11]. The solid curve in Fig. 3.2 for the Au (100)-(hex) surface was measured during anodic potential scan from −0.4 to 0.6 V (SCE). The value of $E_{pzc} = 0.27$ V (SCE) for the Au (100)-(hex) surface is obtained from the location of the differential capacity minimum in the solid curve. The dotted curve in Fig. 3.2 for the Au (100)-(1 × 1) surface was measured during cathodic potential scan from 0.6 to −0.4 V (SCE) after lifting of the Au (100)-(hex) surface. The value of $E_{pzc} = 0.03$ V (SCE) for the Au (100)-(1 × 1) surface is also obtained from the location of the differential capacity minimum in the dotted curve. The difference of 0.24 V in E_{pzc} between the Au (100)-hex and -(1 × 1) surfaces is consistent with that obtained by Kolb [1, 2].

As explained in Sect. 1.9 of Chap. 1, the Lippmann equation: $\left(\frac{\partial \gamma}{\partial E}\right)_{T,\mu_i} = -q$ can be used for a solid metal electrode as well as a liquid metal electrode such as mercury. Consequently, the potential dependence of surface tension $\gamma(E)$, i.e., the electrocapillary curve (γ vs. E) for the Au (100) electrode surfaces, can be calculated from the double integration of differential capacity $c(E)$ with respect to potential E: $\gamma(E) = \gamma_{pzc} - \iint_{E_{pzc}}^{E} c(E)\mathrm{d}E$, where γ_{pzc} is the surface tension at E_{pzc}. Although the absolute value of γ_{pzc} for the Au (100) electrode surfaces cannot be obtained from experiments, the theoretical value of γ for zero surface charge may be available for γ_{pzc}. Ibach et al. [12] calculated $\gamma(E)$ for the Au (100)-(hex) and -(1 × 1) surfaces from the double integration of the differential capacity data in Fig. 3.2 by employing the value of $\gamma_{pzc} = 1.25$ J m^{-2} obtained with first-principles calculation [13–16].

However, the value of $\gamma_{pzc} = 1.25$ J m^{-2} is rather pertinent to the Au (111)-(1 × 1) surface than the Au (100)-(1 × 1) surface [13, 14]. The surface energy of the Au (100)-(1 × 1) surface is larger than that of the Au (111)-(1 × 1) surface since the atomic density of the former is lower than that of the latter. The value of $\gamma_{pzc} = 1.44$ J m^{-2} is preferable for the Au (100)-(1 × 1) surface [15]. Moreover, the value of γ_{pzc} of the Au (100)-(hex) surface has not been theoretically obtained but it should be less than that of the Au (100)-(1 × 1) surface since the Au (100)-(1 × 1) surface has the low atomic density as compared to the Au (100)-(hex) surface. The difference in γ_{pzc} between Au (100)-(1 × 1) and -(hex) surfaces is estimated to be 0.02–0.04 J m^{-2} [17, 18].

Figure 3.3 shows the electrocapillary (γ vs. E) curves of the Au (100)-(hex) and -(1 × 1) surfaces calculated from the differential capacity data (Fig. 3.2). In Fig. 3.3, it is reminded that $\gamma_{pzc} = 1.44$ J m^{-2} is employed for the Au (100)-(1 × 1) surface in place of $\gamma_{pzc} = 1.25$ J m^{-2}. Furthermore, the value of $\Delta\gamma_{pzc} = -0.035$ J m^{-2} is chosen for the difference in γ_{pzc} between Au (100)-(hex) and -(1 × 1) surfaces, so that the electrocapillary curves of the Au (100)-(hex) and -(1 × 1) surfaces intersect at 0.55 V (SCE) corresponding to the critical potential for lifting in 0.01 M $HClO_4$ solution, and $\gamma(E)$ for the Au (100)-(1 × 1) surface becomes less than that for the Au (100)-(hex) surface at potentials more positive than 0.55 V (SCE), which is consistent with the results obtained by Santos and Schmickler [18]. The necessary condition for lifting of reconstruction is that the Au (100)-(1 × 1) surface becomes more thermodynamically stable than that the Au (100)-(hex) surface.

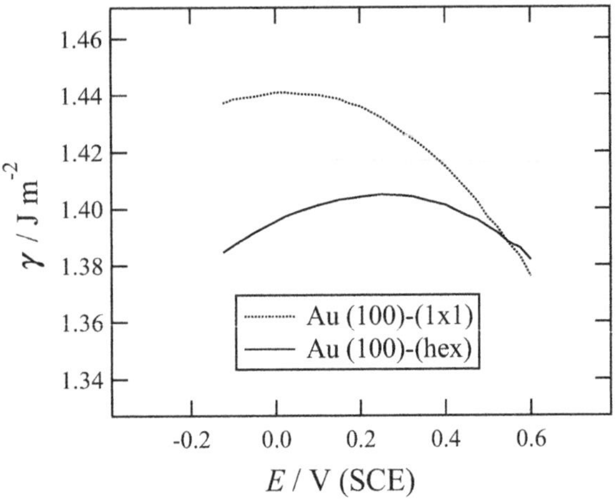

Fig. 3.3 Electrocapillary (γ vs. E) curves of the Au (100)-(hex) and -(1 × 1) surfaces calculated from the differential capacity data (Fig. 3.2). Reprinted from [18], Copyright 2004, with permission from Elsevier

Figure 3.4 shows the surface stress versus potential (g vs. E) curves of the Au (100) surfaces in 0.1 M $HClO_4$ solution measured by a cantilever bending method during continuous anodic polarization from −0.14 to 0.96 V (SCE) at a potential scan rate of 40 mV s^{-1} [8, 9]. The solid curve in Fig. 3.4 was measured on a clean

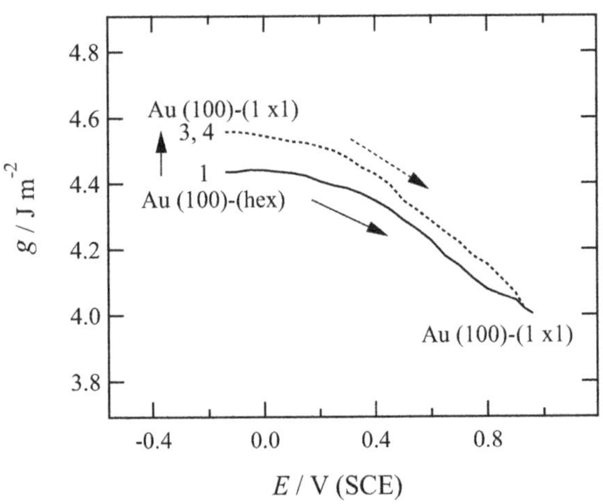

Fig. 3.4 Surface stress versus potential (g vs. E) curves of the Au (100) surfaces in 0.1 M $HClO_4$ solution measured by a cantilever bending method during continuous anodic polarization from −0.14 to 0.96 V (SCE) at a potential scan rate of 40 mV s^{-1} [8, 9]. The numeral in Fig. 3.4 represents the cycle number. Reprinted from [8], Copyright 1997, with permission from Elsevier

prepared Au (100)-(hex) surface during the first anodic potential scan. The Au (100)-(hex) surface has a one-dimensional stripe structure with every sixth row of Au atoms along a [110] direction residing on top of the substrate Au atoms rather than in the fourfold hollow site [17, 19]. A statistical analysis of the virgin Au specimens by scanning tunnel microscopy (STM) indicated that about 50% of the surface area is reconstructed [8, 9]. The change and disappearance of the stripe patterns during the first anodic potential scan revealed that the reconstructed Au (100)-(hex) surface begins to dissolve at 0.26 V (SCE) and the reconstruction is lifted at 0.51 V (SCE). Furthermore, it was observed that the Au atoms removed by lifting form islands on top of the surface.

The curve shifts upward during the second anodic potential scan after the potential returned to −0.14 V (SCE) and then attains to the steady state represented by the dotted curve after 3 or 4 cycles. After 3 or 4 cycles, the surface displayed the unreconstructed Au (100)-(1 × 1) structure even at −0.14 V (SCE). The potentiodynamic STM images [17] indicated that the Au (100)-(hex) surface in 0.1 M $HClO_4$ solution is restored at potentials more negative than −0.25 V (SCE), supporting that the Au (100)-(1 × 1) surface is stable at −0.14 V (SCE). As shown in Fig. 3.4, both solid and dotted curves coincide at about 0.9 V (SCE) where the reconstruction is completely rifted. In Fig. 3.4, the value of $g = 4.57\ \mathrm{J\ m^{-2}}$ obtained by first-principles calculations [15] is employed for the Au (100)-(1 × 1) surface at −0.14 V (SCE) since the absolute value of g cannot be measured by a cantilever bending method. The difference in g between the Au (100)-(hex) and -(1 × 1) surfaces at – 0.14 V (SCE) in Fig. 3.4 is $\Delta g = -0.123\ \mathrm{J\ m^{-2}}$, which is corrected to $\Delta g = -0.25\ \mathrm{J\ m^{-2}}$ for the fully reconstructed Au (100)-(hex) surface since about 50% of the total surface area is reconstructed. As a result, the absolute value of $g = 4.32\ \mathrm{J\ m^{-2}}$ is obtained for the Au (100)-(hex) surface. It is noteworthy that the difference of $\Delta g = -0.25\ \mathrm{J\ m^{-2}}$ between the Au (100)-(hex) and -(1 × 1) surfaces at – 0.14 V (SCE) is about 7 times as much as the difference of $\Delta\gamma_{pzc} = -0.035\ \mathrm{J\ m^{-2}}$ between both surfaces at E_{pzc}.

The positive value of g for a clean metal surface means that the surface is subjected to tensile stress. A clean metal surface has always tensile stress ($g > 0$) since the redistribution of the electronic charge which does not participate in bonding increases the charge density between the surface atoms to reduce the bond distance between the surface atoms in addition to the contraction of the interlayer distance due to the enhancement of the charge between the first and second layers [7, 20]. When the Au (100)-(1 × 1) surface is reconstructed to the Au (100)-(hex) surface, the tensile stress is spontaneously released since the reconstructed Au (100)-(hex) surface has the high atom density as compared to the unreconstructed Au (100)-(1 × 1) surface. In Fig. 3.4, the value of g for the Au (100)-(1 × 1) surface decreases monotonously with increasing potential from −0.14 to 0.96 V (SCE). The potential dependence of g is named "potential-induced surface stress" [20], which does not resemble the corresponding electrocapillary curve in Fig. 3.3. For the Au (100)-(1 × 1) surface, the absolute value of g at $E_{pzc} = 0.03$ V (SCE) is three times as much as that of γ_{pzc}. In addition, the potential dependence of g for the Au (100)-(1 × 1) surface is significantly larger than the potential dependence of γ in the potential region more positive than E_{pzc}.

It is known that the surface stress changes with adsorption of electrolyte anions, which is named "adsorbate-induced surface stress" [21, 22]. The adsorption of electrolyte anions promotes the rifting of reconstruction [1, 2]. The adsorption of ClO_4^- ions on the Au (100) electrode surfaces proceeds at potentials more positive than E_{pzc} [1, 2, 10]. The surface coverage of adsorbed ClO_4^- ions increases with increasing potential, which enhances the repulsive interaction between adsorbed ClO_4^- ions through their dipole moments formed by the charge and the corresponding images. In the case where the adsorbate–adsorbate interaction is repulsive, g as well as γ may decrease with increasing potential, i.e., with increasing surface coverage of adsorbed ClO_4^- ions. According to the calculation made by Schmickler and Leiva [23], using jellium model of metal surface and statistical mechanics of two-dimensional lattice gas, the potential dependence of g is much larger than that of γ. Moreover, their calculation results [23] indicated that the surface concentration of electrons has a direct effect on g while the repulsive adsorbate–adsorbate interaction has an indirect effect on g. The above theoretical results were also supported by Ibach [24].

3.2.2 Au (111) Surface

The rifting of the reconstructed Au (111)-$\left(\sqrt{3} \times 22\right)$ surface as well as the reconstructed Au (100)-(hex) surface has been confirmed by anodic polarization at potentials more positive than the critical potential [8, 9]. The critical potential for lifting of the Au (111)-$\left(\sqrt{3} \times 22\right)$ surface in 0.01 M $HClO_4$ solution is about 0.4 V (SCE), which is more negative by about 0.15 V than that for lifting of the Au (100)-(hex) surface in the same solution. The values of E_{pzc} determined from the differential capacity minima in the c versus E curves measured in 0.01 M $HClO_4$ solution [18] are $E_{pzc} = 0.33$ and 0.23 V (SCE) for the Au (111)-$\left(\sqrt{3} \times 22\right)$ and -(1 × 1) surfaces, respectively, which are consistent with the results obtained by Kolb [1, 2].

Figure 3.5 shows the electrocapillary (γ vs. E) curves of the Au (111)-$\left(\sqrt{3} \times 22\right)$ and -(1 × 1) surfaces calculated from the differential capacity data by Santos and Schmickler [18]. The symbol of Au (111)-(rec) in Fig. 3.5 is an abbreviation of the Au (111)-$\left(\sqrt{3} \times 22\right)$ surface. It is reminded in Fig. 3.5 that the absolute value of $\gamma_{pzc} = 1.25\ \mathrm{J\ m^{-2}}$ obtained by first-principles calculations [13, 14] is employed for the Au (111)-(1 × 1) surface in place of γ_{pzc} which was referred to zero for the Au (111)-$\left(\sqrt{3} \times 22\right)$ surface in their original figure [18]. Furthermore, the value of $\Delta\gamma_{pzc} = -4.7 \times 10^{-3}\ \mathrm{J\ m^{-2}}$ is chosen for the difference in γ_{pzc} between the Au (111)-$\left(\sqrt{3} \times 22\right)$ and -(1 × 1) surfaces, so that the electrocapillary curves of the Au (111)-(1 × 1) and -$\left(\sqrt{3} \times 22\right)$ surfaces intersect at 0.4 V (SCE) corresponding to the critical potential for lifting of the reconstruction in 0.01 M $HClO_4$ solution, and thereby $\gamma(E)$ for the Au (111)-(1 × 1) surface becomes less than that for the

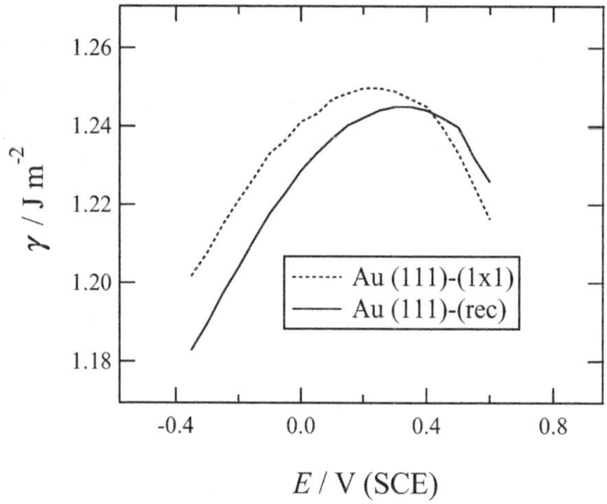

Fig. 3.5 Electrocapillary (γ vs. E) curves of the Au (111)-$\left(\sqrt{3} \times 22\right)$ and -(1 × 1) surfaces calculated by Santos and Schmickler from the differential capacity data [18]. The symbol of Au (111)-(rec) in Fig. 3.5 is an abbreviation of the Au (111)-$\left(\sqrt{3} \times 22\right)$ surface. Reprinted from [18], Copyright 2004, with permission from Elsevier

Au (111)-$\left(\sqrt{3} \times 22\right)$ surface at potentials more positive than 0.4 V (SCE). It is noteworthy that $\gamma_{pzc} = 1.41$ J m^{-2} obtained for the Au (100)-(hex) surface (in Fig. 3.3) is significantly larger than $\gamma_{pzc} = 1.25$ J m^{-2} for the Au (111)-(1 × 1) surface in spite of the similar hexagonal structure. The Au atoms residing on top of the substrate Au atoms in the one-dimensional stripe structure of the Au (100)-(hex) surface [17, 19] are energetically unstable as compared to the Au atoms in the Au (111)-(1 × 1) surface, which may lead to the difference in γ_{pzc} between the Au (100)-(hex) and Au (111)-(1 × 1) surfaces.

Figure 3.6 shows the surface stress versus potential (g vs. E) curves of the Au (111) surfaces in 0.1 M $HClO_4$ solution measured by a cantilever bending method during continuous anodic polarization from −0.14 to 0.96 V (SCE) at a potential scan rate of 40 mV s^{-1} [8, 9]. The solid curve in Fig. 3.6 was measured on a clean prepared Au (111)-$\left(\sqrt{3} \times 22\right)$ surface during the first anodic potential scan. The STM image of the Au (111)-$\left(\sqrt{3} \times 22\right)$ surface displayed the chevron domain pattern characteristic for the reconstructed surface [25]. A statistical analysis of the virgin Au specimens by STM indicated that about 70% of surface area is reconstructed [8, 9]. The domain stripes of the virgin Au surfaces disappeared with the first anodic potential scan to show the dissolution of the reconstructed Au (111)-$\left(\sqrt{3} \times 22\right)$ surface. The typical chevron pattern was no longer visible at 0.96 V (SCE), indicating that the reconstruction is completely rifted. The curve shifts upward with the second anodic potential scan after the potential returned to −0.14 V (SCE) and attains to the steady state

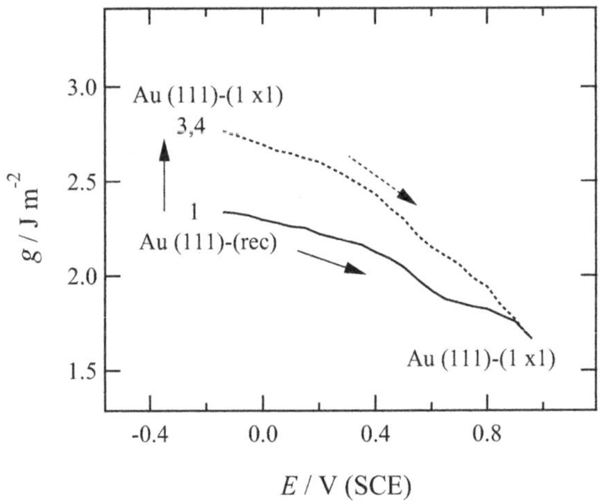

Fig. 3.6 Surface stress versus potential (g vs. E) curves of the Au (111) surfaces in 0.1 M $HClO_4$ solution measured by a cantilever bending method during continuous anodic polarization from −0.14 to 0.96 V (SCE) at a potential scan rate of 40 mV s^{-1} [8, 9]. The numeral in Fig. 3.6 represents the cycle number. The symbol of Au (111)-(rec) in Fig. 3.6 is an abbreviation of the Au (111)-$\left(\sqrt{3} \times 22\right)$ surface. Reprinted from [8], Copyright 1997, with permission from Elsevier

represented by the dotted curve after 3 or 4 cycles. After 3 or 4 cycles, the surface displayed the unreconstructed Au (111)-(1 × 1) structure even at −0.14 V (SCE).

As shown in Fig. 3.6, both solid and dotted curves coincide at about 0.9 V (SCE) where the reconstruction is completely rifted. In Fig. 3.6, the absolute value of $g = 2.77$ J m^{-2} obtained by first-principles calculations [13, 14] is employed for the Au (111)-(1 × 1) surface at −0.14 V (SCE). The difference in g between the Au (111)-$\left(\sqrt{3} \times 22\right)$ and -(1 × 1) surfaces at −0.14 V (SCE) in Fig. 3.6 is $\Delta g = -0.42$ J m^{-2}, which is corrected to $\Delta g = -0.60$ J m^{-2} for the fully reconstructed Au (111)-$\left(\sqrt{3} \times 22\right)$. Consequently, the absolute value of $g = 2.17$ J m^{-2} is obtained for the Au (111)-$\left(\sqrt{3} \times 22\right)$ surface. The absolute value of $g = 4.32$ J m^{-2} obtained for the Au (100)-(hex) surface at −0.14 V (SCE) is significantly larger than $g = 2.77$ J m^{-2} for the Au (111)-(1 × 1) surface at −0.14 V (SCE) in spite of the similar hexagonal structure. Moreover, the difference of $\Delta g = 1.55$ Jm^{-2} between the Au (100)-(hex) and Au (111)-(1 × 1) surfaces at −0.14 V (SCE) is larger by one order of magnitude than the difference of $\Delta \gamma_{pzc} = 0.16$ J m^{-2} between the Au (100)-(hex) and Au (111)-(1 × 1) surfaces at E_{pzc}.

The increase in electronic charge density at the metal surface provides the increase in tensile stress due to the enhancement of attractive force between surface atoms [7, 20]. The corrugated quasi-hexagonal close-packed structure with parallel stripes of Au atoms for the Au (100)-(hex) surface [17, 19] may have the high electronic charge density as compared to the planar hexagonal close-packed structure of the

Au (111)-(1 × 1) surface, and thereby the Au (100)-(hex) surface is subjected to the large tensile stress as compared to the Au (111)-(1 × 1) surface. Besides, it is noticed that the difference in g (i.e., $\Delta g = -0.60\ \mathrm{J\,m^{-2}}$) between the Au (111)-$\left(\sqrt{3} \times 22\right)$ and -(1 × 1) surfaces at −0.14 V (SCE) is significantly larger than that (i.e., $\Delta g = -0.25\ \mathrm{J\,m^{-2}}$) between the Au (100)-(hex) and -(1 × 1) surfaces at −0.14 V (SCE), suggesting that the reconstruction on Au (111) is brought by the large relaxation of tensile stress toward compressive direction.

3.2.3 Roles of Surface Stress in Surface Reconstruction

The relaxation of surface stress toward compressive direction has been measured at the reconstruction of the Au (111) and (100) surfaces [8, 9]. The reconstruction may be driven by the relaxation of surface stress toward compressive direction if the gain in elastic energy due to the relaxation of surface stress is large enough to induce the reconstruction. The relaxation of the surface stress ($\Delta g = -0.60\ \mathrm{J\,m^{-2}}$) at the reconstruction of the Au (111) surface is larger than that ($\Delta g = -0.25\ \mathrm{J\,m^{-2}}$) at the reconstruction of the Au (100) surface. Ibach et al. [8, 9, 20] employed the continuum model to estimate the relaxation of surface stress from the magnitude of uniaxial compression in reconstruction of the Au (111) surface. By considering that the reconstruction induces one-dimensional compression in the surface layer, the change of bulk stress $\Delta\sigma_b$ in the surface layer is given by

$$\Delta\sigma_b = \frac{E_{(111)}\varepsilon}{\left(1 - \nu_{(111)}^2\right)} \tag{3.1}$$

where ε is the one-dimensional strain caused by the reconstruction, and $E_{(111)}$ and $\nu_{(111)}$ are Young's modulus and Poisson's ratio for the Au (111) electrode, respectively. Assuming that on average, one half of the domain orientations point to a particular direction, the bulk stress averaged over all domain orientations corresponds to one half of the stress change calculated with Eq. (3.1) [8].

If the reconstructed surface layer is regarded as a free-standing elastic film, the surface stress relaxation Δg is formulated by

$$\Delta g = \frac{\Delta\sigma_b d_{(111)}}{2} = \frac{E_{(111)}\varepsilon d_{(111)}}{2\left(1 - \nu_{(111)}^2\right)} \tag{3.2}$$

where $d_{(111)}$ is the thickness of the surface atomic layer. The values of $E_{(111)} =$ 81.3 GPa and $\nu_{(111)} = 0.57$ can be obtained from the values of elastic compliances ($S_{11} = 2.34 \times 10^{-11}\ \mathrm{Pa^{-1}}$, $S_{12} = -1.07 \times 10^{-11}\ \mathrm{Pa^{-1}}$, and $S_{44} = 2.38 \times 10^{-11}\ \mathrm{Pa^{-1}}$) for the Au (111) electrode (see Eqs. (4.18) and (4.19) in Sect. 4.3.1 of Chap. 4). The value of $d_{(111)} = 0.235$ nm is estimated from the lattice constant $a = 0.4786$ nm of

Au. Furthermore, the value of $\varepsilon = -0.042$ is estimated from the uniaxial compression along the [110] direction in the Au (111)-$\left(\sqrt{3} \times 22\right)$ surface. As a result, $\Delta g = -0.59$ J m^{-2} is eventually calculated by substituting the above-estimated values into Eq. (3.2), which is in good agreement with the experimental data ($\Delta g = -0.60$ J m^{-2}), supporting that the reconstruction of the Au (111)-(1 × 1) surface is driven by the surface stress relaxation [8].

In contrast to the reconstruction of the Au (111)-(1 × 1) surface, the surface stress relaxation involved in the reconstruction of the Au (100)-(1 × 1) surface is significantly small ($\Delta g = -0.25$ J m^{-2}). The gain in elastic energy due to the small surface stress relaxation would not be sufficient to compensate for the extra energy to put the surface atoms in the less stable sites (i.e., on top of the substrate atoms). Therefore, it is deduced that the surface stress relaxation is not the driving force for the reconstruction of the Au (100)-(1 × 1) surface. The reconstruction of the Au (100)-(1 × 1) surface would be driven by the increase in number of the nearest-neighbor atom due to the formation of the Au (100)-(hex) surface [8].

3.3 Adsorption of Electrolyte Anions

3.3.1 Au (111) Electrode in Acid Solutions Containing ClO_4^-, SO_4^{2-}, and Cl^-

Haiss et al. [22] measured the voltammogram and the changes in surface stress for a (111)-textured Au thin-film electrode in acid solutions containing ClO_4^-, SO_4^{2-}, and Cl^- ions. A cantilever in STM [21, 26] on which a (111)-textured Au thin film was evaporated is used for the measurement of the surface stress change. During the measurement, the (111)-textured Au (111) thin-film surface was kept in the unreconstructed state by using somewhat higher potential scan rate (0.20 V s^{-1}) because of a slow reconstruction process.

Figure 3.7 shows (a) the voltammogram of the (111)-textured Au thin-film electrode in 1.0 M H_2SO_4 solution, (b) the concomitant changes in surface stress Δg, and (c) the first derivative $\frac{\partial g}{\partial E}$ of Δg with respect to potential [22]. The small anodic and cathodic current peaks at about 0.7 V (SCE) in Fig. 3.7a are associated with the formation and disappearance of an ordered sulfate overlayer [27]. In Fig. 3.7b, the value of Δg is referred to zero at −0.2 V (SCE) and the surface stress shifts toward compressive direction as the potential becomes more positive, i.e., with increasing surface coverage of adsorbed sulfate ions. It is remarked that the $\frac{\partial g}{\partial E}$ versus E curve in Fig. 3.7c has a close resemblance to the voltammogram in Fig. 3.7a. Particularly, the potentials (0.4 and 0.3 V (SCE), respectively) at anodic current and cathodic current peaks in Fig. 3.7a coincide with those at the corresponding $\frac{\partial g}{\partial E}$ peaks in Fig. 3.7c. Therefore, the following relationship holds approximately:

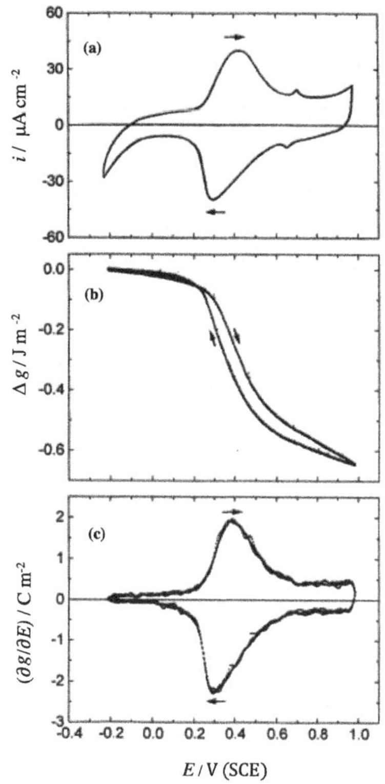

Fig. 3.7 **a** Voltammogram of the (111)-textured Au thin-film electrode in 1.0 M H_2SO_4 solution, **b** the concomitant changes in surface stress Δg, and **c** the first derivative $\frac{\partial g}{\partial E}$ of Δg with respect to potential [22]. Reprinted from [22], Copyright 1998, with permission from Elsevier

$$\frac{\partial g}{\partial E} \approx ki = k\frac{\partial q}{\partial t} = kv\frac{\partial q}{\partial E} \tag{3.3}$$

where k is the proportional constant, q is the surface charge density of the electrode, and v is the potential scan rate (V s^{-1}). Equation (3.3) implies that the surface stress changes linearly with the surface charge density.

Haiss et al. [22] found for the first time the linear dependence of changes in surface stress Δg on surface charge density q as shown in Fig. 3.8 for the (111)-textured Au thin-film electrode in (a) 0.1 M $HClO_4$ (between 0.1 and 0.65 V (SCE) at 0.2 V s^{-1}), (b) 1 M H_2SO_4 (between 0.75 and 0.95 V(SCE) at 0.2 V s^{-1}), and (c) 0.1 M $HClO_4$ + 5 × 10^{-3} M CsCl (between 0.1 and 0.825 V(SCE) at 0.5 V s^{-1}). It is striking that the slope of the linear dependence (−0.91 V for ClO_4^- < −0.85 V for SO_4^{2-} < −0.67 V for Cl^-) changes in accordance with the strength of specific adsorption for electrolyte anions: Cl^- > SO_4^{2-} > ClO_4^-. It is also noted that the dimension of the slope is "voltage." The potential of zero charge E_{pzc} = 0.25 V (SCE) for the Au (111) electrode in perchloric acid [22] is chosen for the determination of q (in the abscissa Fig. 3.8) from the integration of i although E_{pzc} shifts depending on the strength of specific adsorption [28]. Nevertheless, the slopes in Fig. 3.8 do not change with

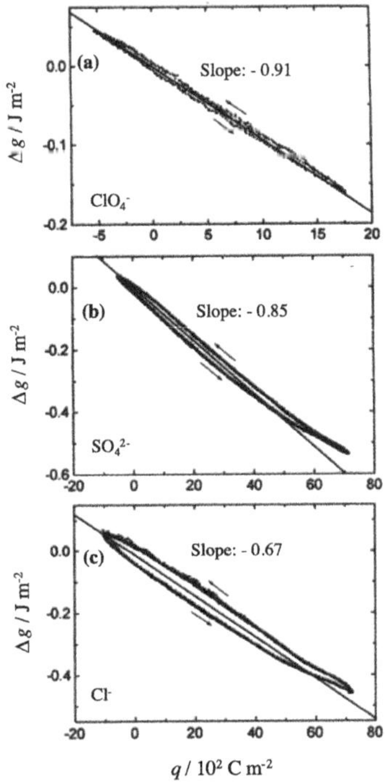

Fig. 3.8 Dependence of changes in surface stress Δg on surface charge density q for the (111)-textured Au (111) thin-film electrode in **a** 0.1 M $HClO_4$ (between 0.1 and 0.65 V (SCE) at 0.2 V s^{-1}), **b** 1 M H_2SO_4 (between 0.75 and 0.95 V(SCE) at 0.2 V s^{-1}), and **c** 0.1 M $HClO_4$ + 5 × 10^{-3} M CsCl (between 0.1 and 0.825 V(SCE) at 0.5 V s^{-1}) [22]. Reprinted from [22], Copyright 1998, with permission from Elsevier

the position of E_{pzc} since the linear relationship between Δg and q moves only in parallel with the abscissa of Fig. 3.8 in response to the shift of E_{pzc}.

In contrast to the linear dependence of Δg on q, the dependence of changes in surface tension $\Delta\gamma$ on q is quadratic (or parabolic) as shown in Fig. 3.9 which is transformed from the electrocapillary curves ($\Delta\gamma$ vs. E) of the Au (111) electrode in 0.1 M $HClO_4$ solutions with and without 5 × 10^{-3} M K_2SO_4 [29]. In Fig. 3.9, the value of $\Delta\gamma$ is referred to zero at $q = 0$ C m^{-2} corresponding to E_{pzc}. If the differential capacity of electric double layer c is potential-independent, the parabolic dependence of $\Delta\gamma$ on E or q can be derived from the Lippmann equation [30]: $\Delta\gamma = \gamma - \gamma_{pzc} = -\frac{(E-E_{pzc})^2}{2c} = -\frac{q^2}{2c}$, where $\Delta\gamma$ is referred to zero at E_{pzc} (see Eq. (1.126) in Sect. 1.9 of Chap. 1). The comparison between Figs. 3.8b and 3.9 for the adsorption of SO_4^{2-} in the same range of $q = 0$ ~ 0.6 C m^{-2} indicates that the dependence of Δg on q is larger by a factor of about 2.5 than the dependence of $\Delta\gamma$ on q.

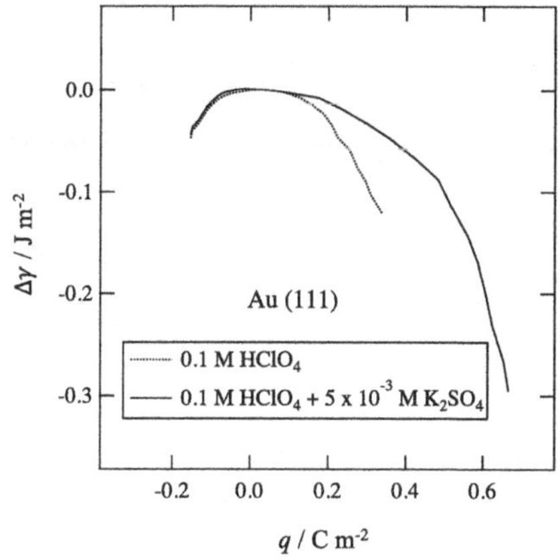

Fig. 3.9 Dependence of changes in surface tension $\Delta\gamma$ on surface charge density q, which was transformed from the electrocapillary curves of the Au (111) electrode in 0.1 M $HClO_4$ solutions with and without 5×10^{-3} M K_2SO_4 [29]

3.3.2 Au (111) Electrode in Perchlorate Solution Containing Iodide Ions

Surface X-ray scattering (SXS) study [31] of the structure of iodine electrodeposited on Au (111) in 0.01 M–0.1 M KI solutions indicated that a $\left(p \times \sqrt{3}\right)$ centered-rectangular phase is formed in the low potential region, while a rotated-hexagonal phase is formed in the high potential region. In both phases, the structures are compressed with increasing potential, i.e., subjected to electro-compression. Ueno and Seo [32] measured the changes in surface stress of a (111)-textured Au (111) thin-film electrode in perchlorate solution containing iodide ions by using a cantilever bending method. The terminology of surface energy γ^s was used in their original paper [32] in place of surface stress g which is a surface thermodynamic quantity measured by a cantilever bending method. Furthermore, the refractive index of solution n_s was not taken into consideration for determination of the curvature radius of the cantilever beam electrode in solution. Nevertheless, provided that the curvature radius is corrected by the refractive index of the solution $n_s = 1.34$, the changes in surface stress can be accurately obtained from their original data.

Figure 3.10 shows (a) the cyclic voltammogram and (b) the Δg versus E curve (after the correction of $n_s = 1.34$ was made) of the (111)-textured Au (111) thin-film electrode measured at a potential scan rate of 20 mV s^{-1} in 0.1 M $NaClO_4$ solution containing 3×10^{-3} M NaI [32]. The cathodic limit and anodic limit potentials are set at −0.80 and 0.42 V (SHE), respectively. The cathodic current flows up to −0.60 V in the anodic potential scan from −0.80 V(SHE), which may be due to the contribution of hydrogen evolution. The cyclic voltammogram for the Au (111) electrode in 10^{-2} M KI solution [31] also indicated that the cathodic current flows up to −0.35 V in the anodic potential scan from −0.50 V (SHE). In Fig. 3.10a, the

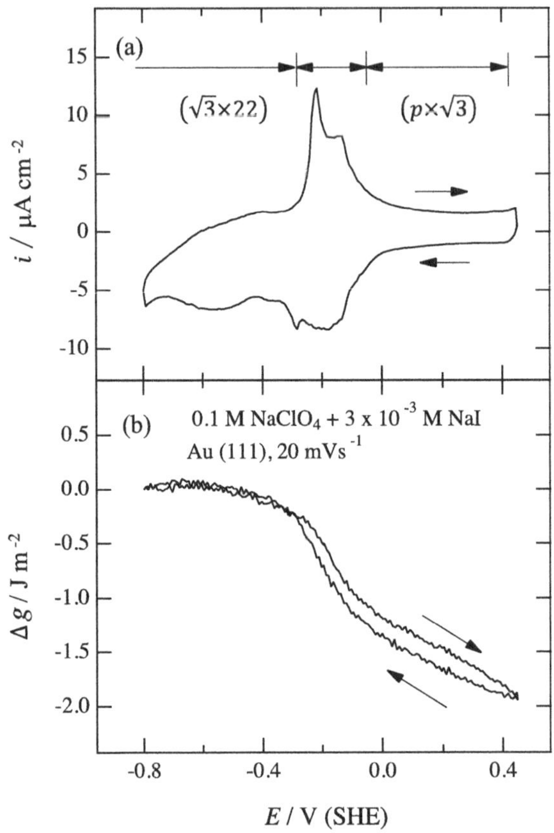

Fig. 3.10 **a** Cyclic voltammogram and **b** the Δg versus E curve (after the correction of $n_s = 1.34$ was made) of the (111)-textured Au (111) thin-film electrode measured at a potential scan rate of 20 mV s^{-1} in 0.1 M $NaClO_4$ solution containing 3×10^{-3} M NaI [32]. Modified from [32] with permission from The Electrochemical Society

sharp anodic current peak observed at −0.215 V (SHE) in the anodic potential scan is caused by the lifting of the reconstructed Au (111)-$\left(\sqrt{3} \times 22\right)$ surface and the broad reversible current peaks centered at about −0.15 V (SHE) correspond to the adsorption/desorption of iodine, which are consistent with the features of the cyclic voltammograms measured in 10^{-2} M KI solution [31] and in 0.1 M $NaClO_4$ solution containing 5×10^{-3} M NaI [33].

According to Ocko et al. [31], the $\left(p \times \sqrt{3}\right)$-iodine phase on Au (111) surface in 10^{-2} M KI solution is stable in the potential region between −0.08 and 0.40 V (SHE). The shift in equilibrium potential of iodine adsorption due to the concentration of iodide ions in solution can be corrected by using the Nernst equation. It is deduced that the $\left(p \times \sqrt{3}\right)$-iodine phase on Au (111) surface in 0.1 M $NaClO_4$ solution containing 3×10^{-3} M NaI is stable in the potential region between −0.06 and 0.42 V (SHE). Small reversible current spikes are usually observed in the cyclic voltammogram at the potential where the $\left(p \times \sqrt{3}\right)$-iodine phase is transformed to the rotated-hexagonal phase [31, 33]. A small rise of anodic current at 0.42 V (SHE) in the cyclic voltammogram of Fig. 3.10a may be associated with the transformation from $\left(p \times \sqrt{3}\right)$ to rotated-hexagonal phase.

The value of Δg is referred to zero at −0.80 V (SHE) in the Δg versus E curve of Fig. 3.10b. The surface stress keeps almost constant in the potential region between −0.80 and −0.40 V (SHE) where the reconstructed Au (111)-$\left(\sqrt{3} \times 22\right)$ surface is stable. The value of Δg decreases rapidly toward compressive direction in the anodic potential scan from −0.3 to −0.1 V (SHE) where the lifting of the reconstruction followed by the adsorption of iodine takes place. The decrease in Δg toward compressive direction continues, keeping an almost linear relationship between Δg and E in the anodic potential scan from −0.1 to 0.42 V (SHE) where the $\left(p \times \sqrt{3}\right)$-iodine phase is stable. In the cathodic potential scan from 0.42 to −0.80 V (SHE), the reversed changes in surface stress with a small hysteresis are observed.

According to the SXS study [31], the nearest-neighbor distance a_{nn} between iodine atoms in the $\left(p \times \sqrt{3}\right)$ iodine adlayer on Au (111) electrode in 0.1 M KI solution decreases linearly from 0.462 to 0.432 nm with increasing potential due to the increase in electro-compression. The elastic strain ε in the iodine adlayer can be calculated from the changes in a_{nn}:

$$\varepsilon = \frac{\left(a_{nn} - a_{nn,o}\right)}{a_{nn,o}}, \tag{3.4}$$

where $a_{nn,o}$ is 0.462 nm. The minimum distance of $a_{nn} = 0.432$ is equal to the van der Waals diameter of iodine atom in which ε amounts to −0.065. The minus sign of ε means that the elastic strain is compressive. The changes in surface stress toward compressive direction should be responsible for the changes in elastic strain toward compressive direction. From the potential correction by the Nernst equation, the potential range where a_{nn} changes from 0.462 to 0.432 nm in 0.1 M KI solution corresponds to the potential range of 0.01 to 0.41 V (SHE) in 0.1 M $NaClO_4$ solution containing 3×10^{-3} M NaI. As shown in the Δg versus E curve of Fig. 3.10b, the value of $\Delta(\Delta g) = -0.63$ J m^{-2} is obtained for the change in potential from 0.01 to 0.41 V (SHE). The $\left(p \times \sqrt{3}\right)$ iodine adlayer on Au (111) is subjected to a uniaxial electro-compression along the incommensurate direction.

In the continuum model, if one-dimensionally compressed adlayer is regarded as a free-standing elastic film, the following linear relationship between $\Delta(\Delta g)$ and $\Delta\varepsilon$ holds:

$$\Delta(\Delta g) = Y_I \Delta\varepsilon d, \tag{3.5}$$

where Y_I and d are the biaxial modulus and thickness of the iodine adlayer, respectively. The value of $Y_I = 39.1$ GPa for the $\left(p \times \sqrt{3}\right)$ iodine adlayer on Au (111) is estimated by substituting $\Delta(\Delta g) = -0.63$ J m^{-2}, $\Delta\varepsilon = -0.065$, and $d = 0.248$ nm for iodine adlayer into Eq. (3.5). The biaxial moduli of any iodine adlayers on noble metals have not been reported so far. The value of $Y_I = 39.1$ GPa for the $\left(p \times \sqrt{3}\right)$ iodine adlayer on Au (111) is less than $Y_{Pb(111)} = 58.4$ GPa for the hexagonal close-packed (hcp) underpotential deposition (UPD) layer of Pb on Au (111) and $Y_{Bi} =$ 77.1 GPa for the UPD layer of Bi with the same $\left(p \times \sqrt{3}\right)$ structure on Au (111) (see Sects. 4.3.1 and 4.3.2 of Chap. 4).

3.4 Surface Stress versus Surface Charge Density or Potential versus Surface Elastic Strain

3.4.1 *Surface Stress–Surface Charge Density Coefficient* $\zeta_{g,q}$

The value of $\zeta_{g,q} = -0.91$ V obtained by Haiss et al. [22] from the slope of the linear relationship between Δg and q for the unreconstructed, textured-(111) Au thin-film electrode in 0.1 M $HClO_4$ solution is consistent with that ($\zeta_{g,q} = -0.86$ V) obtained by Ibach [24]. Smetanin et al. [35] measured the changes in surface stress of a (111)-textured Au thin-film electrode in 7×10^{-3} M NaF and 0.01 M $HClO_4$ solutions and found the linear relationship between Δg and q near E_{pzc}. The value of $E_{pzc} = 0.20$ V (SCE) in 0.01 M $HClO_4$ solution determined from the capacity minimum in the measured differential capacity versus potential curve [35] is close to $E_{pzc} = 0.23$ V (SCE) in 0.01 M $HClO_4$ solution reported for the unconstructed Au (111) surface [36], confirming that the (111)-textured Au thin-film electrode used for the measurement of changes in surface stress is unreconstructed. The surface stress–surface charge density coefficients obtained from the linear slopes [35] are $\zeta_{g,q} = -1.95$ V for 7×10^{-3} M NaF solution and -2.0 V for 0.01 M $HClO_4$ solution, which are larger by a factor of two than those obtained by Haiss et al. [22] and by Ibach [24], while they are in good agreement with the results ($\zeta_{g,q} = -1.86$ V) of ab initio calculations for Au (111) in vacuum [37].

It has been reported [38] that $\zeta_{g,q}$ in 0.1 M $HClO_4$ solution for a fresh and clean Au (111) surface is close to -2.0 V. The value of $\zeta_{g,q}$ may be influenced by the contamination of the Au (111) surface. At present, however, the causes for the discrepancy of $\zeta_{g,q}$ between the researchers have not been made clear. The value of $\zeta_{g,q}$ obtained from the slope of the linear relationship between Δg and q is not sensitive to potential since it is an average value over the specified potential region. We discuss the results obtained by a potential-sensitive technique of $\zeta_{g,q}$ in the next pages.

3.4.2 Determination of $\zeta_{g,q}$ by Dynamic Stress Analysis Combined with Electrochemical Impedance Spectroscopy

Surface stress–surface charge density coefficient $\zeta_{g,q}$ can be determined as a function of potential by a dynamic stress analysis (DSA) combined with electrochemical impedance spectroscopy (EIS) using a cantilever bending [39, 40]. The principle of DSA combined with EIS is briefly explained in Sect. 2.3.2 of Chap. 2. In a conventional EIS experiment, the system responds to the application of a sinusoidal potential, $E = E_{dc} + E_o \exp(j\omega t)$, where E_{dc} is a dc potential, E_o is the signal amplitude, and ω is the angular frequency. If E_o is sufficiently small, the current response is linear and is formulated by $i = i_{dc} + i_o \exp[j(\omega t + \psi_e)]$, where ψ_e is the phase angle between current and potential. On the other hand, the corresponding surface stress response is formulated by $g = g_{dc} + g_o \exp[j(\omega t + \psi_s)]$, where ψ_s is the phase angle between g and E. The ac component g_{ac} of the total surface stress is $g_o \exp[j(\omega t + \psi_s)]$.

By the way, the complementary explanation of stress impedance in addition to electrochemical impedance is needed to understand the results obtained by DAS combined with EIS. The electrochemical impedance Z_e is given by [39, 40]:

$$Z_e = \frac{E_o \exp(j\omega t)}{i_o \exp[j(\omega t + \psi_e)]} = \frac{E_o}{i_o} \exp(-j\psi_e). \tag{3.6}$$

The absolute value of Z_e is $|Z_e| = \frac{E_o}{i_o}$. In analogy with the electrochemical impedance, the stress impedance Z_s is formulated by [39, 40]:

$$Z_s = \frac{E_o \exp(j\omega t)}{g_o \exp[j(\omega t + \psi_s)]} = \frac{E_o}{g_o} \exp(-j\psi_s). \tag{3.7}$$

The reciprocal of Z_s is called the stress admittance, Y_s (=Z_s^{-1}). The ac component q_{ac} of the total surface charge density can be obtained by integrating the ac density component $i_o \exp[j(\omega t + \psi_e)]$ with respect to time t:

$$q_{ac} = \frac{i_o \exp[j(\omega t + \psi_e)]}{j\omega} = \frac{E_o \exp(j\omega t)}{j\omega Z_e}. \tag{3.8}$$

The ratio of g_{ac} to q_{ac} is equivalent to $\zeta_{g,q} = \frac{\partial g}{\partial q}$ and is finally derived from Eqs. (3.6), (3.7), and (3.8) as follows [39, 40]:

$$\begin{aligned}\zeta_{g,q} &= \frac{g_{ac}}{q_{ac}} = j\omega \frac{g_o}{i_o} \exp[j(\psi_s - \psi_e)] \\ &= \omega \frac{g_o}{i_o}\left[\cos\left(\psi_s - \psi_e + \frac{\pi}{2}\right) + j\sin\left(\psi_s - \psi_e + \frac{\pi}{2}\right)\right] \\ &= j\omega Y_s Z_e \end{aligned} \tag{3.9}$$

Equation (3.9) means that the value of $\zeta_{g,q}$ can be obtained from the values of Y_s and Z_e. The phase angle of $\zeta_{g,q}$ is equal to the argument $\left(\psi_s - \psi_e + \frac{\pi}{2}\right)$ in Eq. (3.9).

The frequency response of an elastic cantilever to an external driving force depends strongly on the cantilever geometry (length: L, width:b, and thickness:h) and on the fluid in which it is immersed. Provided that the cantilever is an isotropic elastic solid with negligible internal friction and its cross section is uniform over whole length under the geometrical restriction of $L \gg b \gg h$, the resonant frequency at flexural mode in vacuum and in fluid can be theoretically predicted [41]. For a rectangular glass cantilever ($L = 25$ mm, $b = 3$ mm, and $h = 0.1$ mm) used as a substrate, the predicted resonant frequencies at the first mode in vacuum and in solution are 139 and 43 Hz, respectively [39].

Figure 3.11 shows (a) the absolute value $\left|\zeta_{g,q}\right|$ of surface stress–surface charge density coefficient and (b) the phase angle $\psi_{g,q}$ of surface stress relative to surface

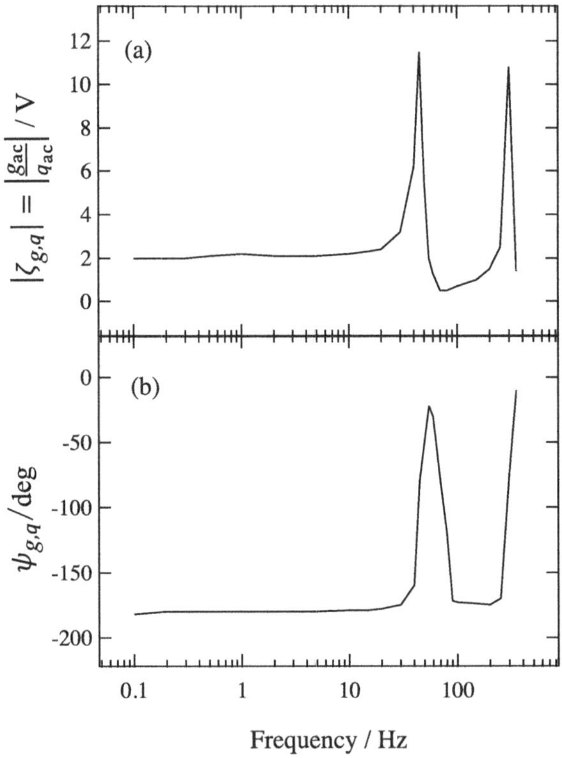

Fig. 3.11 **a** Absolute value $\left|\zeta_{g,q}\right|$ of surface stress–surface charge density coefficient and **b** the phase angle $\psi_{g,q}$ of surface stress relative to surface charge density for the (111)-textured Au (111) thin-film (cantilever) electrode ($L = 24.2$ mm, $b = 3$ mm, and $h = 0.1$ mm) in 0.1 M $HClO_4$ solution, responding to a sinusoidal potential input of ± 50 mV at a steady-state potential of 0.2 V (SSE) where no faradaic process takes place [39]. Reproduced from [39] with permission from The Electrochemical Society

charge density for a (111)-textured Au thin-film (cantilever) electrode ($L = 24.2$ mm, $b = 3$ mm, and $h = 0.1$ mm) in 0.1 M $HClO_4$ solution, responding to a sinusoidal potential input of ±50 mV at a steady-state potential of 0.2 V (SSE) where no faradaic process takes place [39]. A saturated Hg/$HgSO_4$ electrode (SSE) is used as a reference electrode in experiment of Fig. 3.11, which is +0.64 V referred to a standard hydrogen electrode (SHE). The two resonant modes at 46 and 300 Hz are observed from the sharp increases in $\left|\zeta_{g,q}\right|$ and $\psi_{g,q}$, which are consistent with those predicted [39]. In the frequency range of 0.1 to 20 Hz, $\left|\zeta_{g,q}\right|$ keeps constant, taking an average value of 2.0 V, while $\psi_{g,q}$ has a constant value of −180° from which an average value of $\zeta_{g,q} = \frac{g_{ac}}{q_{ac}} = -2.0$ V is obtained in the entire frequency range where the cantilever is non-resonant. It is confirmed [39] that the value of $\left|\zeta_{g,q}\right|$ is not influenced by the amplitude of the sinusoidal potential input in the range of ±5 mV to ±80 mV but the uncertainty in $\left|\zeta_{g,q}\right|$ is larger for the smaller amplitude of signals. The value of $\zeta_{g,q} = -2.0$ V determined by DAS combined with EIS is consistent with those reported for the (111)-textured Au thin-film electrode in NaF and $HClO_4$ solutions by Smetanin et al. [35].

Figure 3.12 shows the $\left|\zeta_{g,q}\right|$ versus E curve for the (111)-textured Au thin-film electrode ($L = 23.6$ mm, $b = 3$ mm, and $h = 0.1$ mm) in 0.1 M $HClO_4$ solution, responding to the input signal amplitude of 50 mV at a frequency of 1 Hz [39]. The values of $\psi_{g,q}$ are described in the arrows on the upper part in the abscissa of Fig. 3.12. In the experiment of Fig. 3.12, after the Au electrode was conditioned at 0.05 V (SSE) for 1 min, E_{dc} is sequentially stepped by 25 mV, starting from 0.15 V anodically to 0.90 V, then from 0.90 V cathodically to −0.60 V, and finally back to

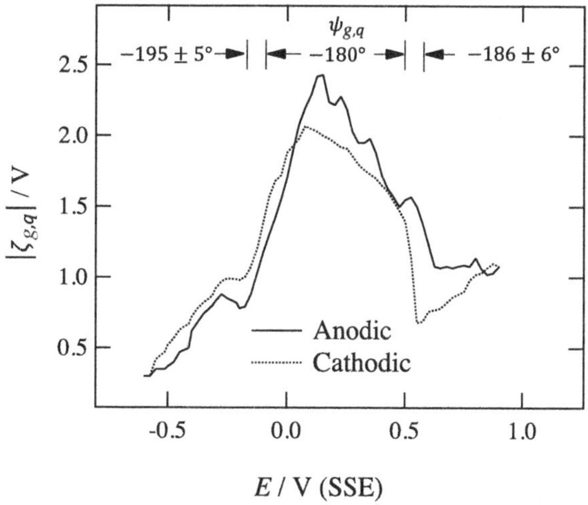

Fig. 3.12 $\left|\zeta_{g,q}\right|$ versus E curve for the (111)-textured Au thin-film electrode ($L = 23.6$ mm, $b = 3$ mm, and $h = 0.1$ mm) in 0.1 M $HClO_4$ solution, responding to the input signal amplitude of 50 mV at a frequency of 1 Hz [39]. Modified from [39] with permission from The Electrochemical Society

0.15 V (SSE). As shown in Fig. 3.12, $|\zeta_{g,q}|$ varies sensitively with potential and takes maxima at about 0.15 V (SSE) in the anodic potential scan and at about 0.1 V (SSE) in the cathodic potential scan. The maximum value of $|\zeta_{g,q}| = 2.4$ V is somewhat higher than that of $|\zeta_{g,q}| = 2.0$ V in the cathodic potential scan. The cyclic voltammogram of the (111)-textured Au thin-film electrode in 0.1 M $HClO_4$ solution [39] exhibited the abrupt increase in anodic current due to oxidation at potentials more positive than 0.65 V (SSE) and the cathodic current peak due to reduction of the oxide film at 0.45 V (SSE) in the cathodic potential scan from 0.90 V (SSE). The hysteresis of $|\zeta_{g,q}|$ in anodic and cathodic potential scans may be associated with the oxide formation/reduction on Au electrode.

The value of $\psi_{g,q}$ keeps constant, taking exactly -180° in the potential region between -0.1 and 0.5 V (SSE), while $\psi_{g,q}$ fluctuates by $\pm 5^{\circ}$ around -195° at potentials more negative than -0.2 V (SSE) and by $\pm 6^{\circ}$ around -186° at potentials more positive than 0.6 V (SSE). The slight deviation of $\psi_{g,q}$ from -180° at potentials more negative than -0.2 V (SSE) and at potentials more positive than 0.6 V (SSE) may result from the reduction of oxygen in solution and from the oxidation of the Au electrode, respectively. Nevertheless, $\psi_{g,q} \approx -180^{\circ}$ means that the sign of $\zeta_{g,q}$ is minus over the entire range of potentials. Consequently, $\zeta_{g,q} = \frac{\partial g}{\partial q}$ takes -2.4 V at about 0.15 V (SSE) in the anodic potential scan and -2.0 V at about 0.1 V (SSE) in the cathodic potential scan. The potential dependence of $\zeta_{g,q}$ in Fig. 3.12 is quite similar to the potential dependence of $\zeta_{E,\varepsilon}$ obtained from the potential variation due to cyclic strain by Smetanin et al. [42] (compare Fig. 3.12 with Fig. 3.13b).

3.4.3 Potential–Surface Elastic Strain Coefficient $\zeta_{E,\varepsilon}$

It has been derived by Gokhshtein [34] that surface stress–surface charge density coefficient $\zeta_{g,q} = \frac{\partial g}{\partial q}$ is equivalent to potential–surface elastic strain coefficient $\zeta_{E,\varepsilon} = \frac{\partial E}{\partial \varepsilon}$ (see Eq. (1.124) in Sect. 1.8 of Chap. 1). Smetanin et al. [43] tried to measure $\zeta_{E,\varepsilon}$ from the potential variation of a (111)-textured Au thin-film electrode subjected to cyclic elastic strain under open-circuit condition in 0.01 M $HClO_4$ solution. The experimental value of $\zeta_{E,\varepsilon}$ exhibited a frequency dependence due to Faraday loss current. However, the frequency dependence became negligible beyond 30 Hz. The experimental value of $\zeta_{E,\varepsilon} = -1.83$ V at higher frequency is close to $\zeta_{g,q} = -2.0$ V obtained by a cantilever bending method [35] and is in good agreement with $\zeta_{g,q} = -1.86$ V predicted by ab initio calculations for Au (111) in vacuum [37].

Furthermore, Smetanin et al. [42] developed a new technique for measuring the current elastic strain or potential elastic strain response as a function of potential during cyclic voltammetry. This technique is named "dynamic electro-chemo-mechanical analysis (DECMA)." The instrumentation and experimental setup of DECMA have been briefly described in Sect. 2.4 of Chap. 2. Nevertheless, detailed explanation of its principle is needed for better understanding of the results obtained by DECMA. In DECMA, surface elastic strain ε is changed sinusoidally with time

t:

$$\varepsilon = \varepsilon_o \sin(\omega t), \tag{3.10}$$

where ω is the angular frequency and ε_o is the amplitude of ε. In the case of an ideally polarizable electrode without Faraday process under open-circuit condition, the concomitant potential variation is formulated as follows [42]:

$$E = E_o \sin(\omega t) = \varepsilon_o \left|\zeta_{E,\varepsilon}\right| \sin(\omega t), \tag{3.11}$$

where $\left|\zeta_{E,\varepsilon}\right|$ is the absolute value of $\zeta_{E,\varepsilon}$. Therefore, $\left|\zeta_{E,\varepsilon}\right|$ can be obtained as the amplitude ratio of E to ε, i.e., $\frac{E_o}{\varepsilon_o}$.

If ε is changed sinusoidally at constant potential under the assumption that ε is small and a linear relationship between q and ε holds, the small change of q will be represented by

$$\delta q = \left(\frac{\partial q}{\partial \varepsilon}\right)_E \delta\varepsilon. \tag{3.12}$$

In addition, $\left(\frac{\partial q}{\partial \varepsilon}\right)_E$ can be divided into the two terms:

$$\left(\frac{\partial q}{\partial \varepsilon}\right)_E = -\left(\frac{\partial q}{\partial E}\right)_\varepsilon \left(\frac{\partial E}{\partial \varepsilon}\right)_q = -c\zeta_{E,\varepsilon}, \tag{3.13}$$

where $\left(\frac{\partial q}{\partial E}\right)_\varepsilon = c$ and $\left(\frac{\partial E}{\partial \varepsilon}\right)_q = \zeta_{E,\varepsilon}$. Consequently, the change of q due to cyclic small strain at equilibrium is given by [42]:

$$\delta q = -c\left|\zeta_{E,\varepsilon}\right|\varepsilon_o \sin(\omega t). \tag{3.14}$$

The corresponding variation of current density i is [42]:

$$i = \frac{\mathrm{d}q}{\mathrm{d}t} = i_o \cos(\omega t), \tag{3.15}$$

and

$$i_o = -c\left|\zeta_{E,\varepsilon}\right|\varepsilon_o \omega, \tag{3.16}$$

or equivalently in the case where the electrode surface area is unknown:

$$I = \frac{\mathrm{d}Q}{\mathrm{d}t} = I_o \cos(\omega t), \tag{3.17}$$

and

$$I_o = -C\left|\zeta_{E,\varepsilon}\right|\varepsilon_0\omega, \tag{3.18}$$

where I, Q, and C are net current, net charge, and net capacity of the electrode, respectively.

The experiment for the potential variation due to surface elastic strain needs a sufficiently high strain frequency to avoid the participation of Faraday loss current in q. However, the strain cycles at such high frequency bring phase shifts of the potential and current responses due to the slow transport process of the electrolyte or due to the slow adsorption process of electrolyte anions [42]. At the high-frequency strain cycles, Eq. (3.11) will be modified to

$$E = \varepsilon_0\left|\zeta_{E,\varepsilon}\right|\sin(\omega t - \phi), \tag{3.19}$$

where ϕ is the phase angle relative to the cyclic elastic strain. The potential amplitude E_o in Eq. (3.19) is given by $\varepsilon_0\left|\zeta_{E,\varepsilon}\right|$. Equation (3.17) will be also modified to

$$I = I_o\cos(\omega t - \phi - \psi), \tag{3.20}$$

where ψ is the phase angle relative to the cyclic potential variation.

Since in conventional electrochemical impedance spectroscopy (EIS), potential and current are correlated by the complex electrochemical impedance $Z_e = Z_{re} - jZ_{im}$, I_o and ψ in Eq. (3.20) are given by [42]:

$$I_o = \frac{\varepsilon_0\left|\zeta_{E,\varepsilon}\right|}{|Z_e|}, \tag{3.21}$$

and

$$\tan\psi = \frac{Z_{im}}{Z_{re}}, \tag{3.22}$$

where $|Z_e|$, Z_{re}, and Z_{im} are the absolute value, real component, and imaginary component of Z_e, respectively. In the case where the equivalent circuit in electric double-layer region simplifies to a series *RC* circuit, Z_e can be expressed by [42]:

$$Z_e = r_{sol} - \frac{j}{\omega c_{dl}}, \tag{3.23}$$

where r_{sol} is the solution resistance and c_{dl} is the differential capacity of the electric double layer. In DECMA, the potential variation is imposed by the cyclic elastic strain, and the concomitant current response is controlled by Z_e. Since $E_o = I_o|Z_e|$, $\left|\zeta_{E,\varepsilon}\right|$ is given by

$$\left|\zeta_{E,\varepsilon}\right| = \frac{E_o}{\varepsilon_0} = \frac{I_o|Z_e|}{\varepsilon_0}. \tag{3.24}$$

The value of $|\zeta_{E,\varepsilon}|$ can be eventually obtained from Eq. (3.24) by measuring I_o in the current–strain response (DECMA) experiment (see Fig. 2.16a in Sect. 2.4 of Chap. 2) and by combining $|Z_e|$ determined with separate EIS experiments under identical conditions with the DECMA experiment [42]. The sign of $\zeta_{E,\varepsilon}$ can be determined from the phase angle ψ.

Alternatively, apart from the current–strain response experiment, $\zeta_{E,\varepsilon}$ can be measured directly by the potential–strain response in connection of a delay resistance R_D in series with the working electrode without the need of separate EIS experiment [42] (see Fig. 2.16b in Sect. 2.4 of Chap. 2). The addition of R_D to the uncompensated solution resistance r_{sol} increases the time constant for potential control at the working electrode, which lowers the performance of the potentiostat compensating the potential variation due to the cyclic strain. As a result, the strain-induced potential variation can be measured by a lock-in amplifier connected between the working and reference electrodes. For a typical electrode with a capacity of 100 μF, the addition of $R_D = 50$ k Ω yields the time constant of about 5 s, much longer than the cycle time (e.g., 0.05 s for 20 Hz) of elastic strain but much shorter than the time (e.g., 10^3 s V^{-1} for 1 mV s^{-1}) to complete a sufficiently slow voltammogram [42].

Figure 3.13 shows (a) the cyclic voltammogram and (b) the $\zeta_{E,\varepsilon}$ versus E curve obtained by the potential–strain response at a potential scan rate of 1 mV s^{-1} under

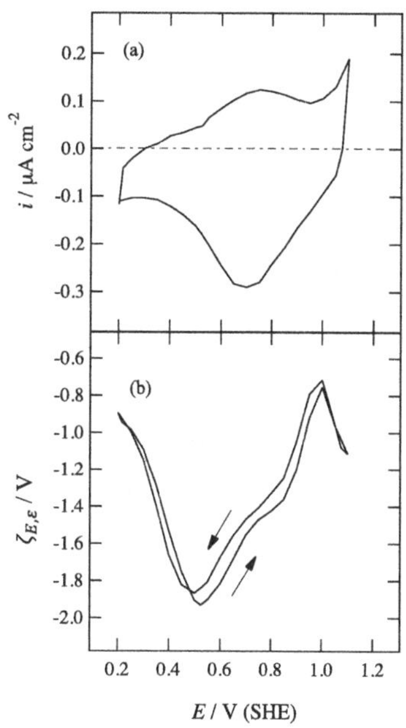

Fig. 3.13 **a** Cyclic voltammogram and **b** the $\zeta_{E,\varepsilon}$ versus E curve obtained by the potential–strain response at a potential scan rate of 1 mV s^{-1} under a strain frequency of 20 Hz with a strain amplitude of $\varepsilon_o = 2 \times 10^{-4}$ for the (111)-textured Au thin-film electrode in 0.01 M $HClO_4$ solution [42]. Reproduced from [42] with permission from the PCCP Owner Societies

a strain frequency of 20 Hz with a strain amplitude of $\varepsilon_o = 2 \times 10^{-4}$ for a (111)-textured Au thin-film electrode in 0.01 M $HClO_4$ solution [42]. The delay resistance $R_D = 50$ k Ω is employed for the experiment of Fig. 3.13, which exhibits only minor difference in cyclic voltammogram at 1 mV s^{-1}. In Fig. 3.13b, the value of $\zeta_{E,\varepsilon}$ with minus sign depends on potential, taking a minimum at $E = 0.53$ V (SHE) although it exhibits a small hysteresis. The potential dependence of $\zeta_{E,\varepsilon}$ obtained by the potential–strain response coincided with that obtained by the current–strain response except for the slight difference in hysteresis [42]. The potential at which $\zeta_{E,\varepsilon}$ takes the minimum is close to the potential of zero charge $E_{pzc} = 0.48$ V (SHE) for the Au (111) electrode in $HClO_4$ solution [36]. The minimum value of $\zeta_{E,\varepsilon} = -1.9 \pm 0.2$ V agrees with that ($\zeta_{E,\varepsilon} = -1.83$ V) obtained by the potential–strain response under open-circuit condition [43].

Moreover, the minimum value of $\zeta_{E,\varepsilon}$ coincides with $\zeta_{g,q} = -2.0$ V obtained at near E_{pzc} by a cantilever bending method [35], proving experimentally that $\zeta_{E,\varepsilon}$ is equivalent to $\zeta_{g,q}$. It has been also found that there are no significant differences between the $\zeta_{E,\varepsilon}$ versus E curves obtained in 10^{-3} M ~ 0.1 M $HClO_4$ and H_2SO_4 solutions, using strain cycles with a frequency of 20 Hz and an amplitude of $\varepsilon_o = 2 \times 10^{-4}$, indicating that $\zeta_{E,\varepsilon}$ or $\zeta_{g,q}$ is essentially independent of the electrolyte anion species and of the solution concentration, which may mean that $\zeta_{E,\varepsilon}$ or $\zeta_{g,q}$ is linked directly to the electronic process of the metal surface such as band-filling by the excess surface charge [37, 42]. The $\zeta_{E,\varepsilon}$ versus E curve in Fig. 3.13b exhibits not only a minimum at 0.53 V but also a shoulder at about 0.8 V and an inversion in slope at about 1.0 V (SHE).

3.4.4 Sign-Reversal of $\zeta_{g,q}$ in the Hydrogen Adsorption/Desorption Region or in the Oxide Formation/Reduction Region

The sign-reversals of the potential dependence of surface stress $\frac{\partial g}{\partial E}$ in the hydrogen adsorption/desorption region and in the oxide formation/reduction region have been qualitatively observed for polycrystalline Pt [44, 45] and Pd [46, 47] foil electrodes in sulfate solutions with different pH values and in perchloric acids with different concentrations of chloride ions by using a piezoelectric technique similar to Gokhshtein's estance method [34]. The following relationship between $\zeta_{g,q}$ and $\frac{\partial g}{\partial E}$ holds:

$$\zeta_{g,q} = \frac{\partial g}{\partial q} = \left(\frac{\partial g}{\partial E}\right)\left(\frac{\partial E}{\partial q}\right) = \frac{1}{c}\left(\frac{\partial g}{\partial E}\right), \tag{3.25}$$

where c is the differential capacity. The sign-reversal of $\frac{\partial g}{\partial E}$ is directly linked to the sign-reversal of $\zeta_{g,q}$ since c depends on potential but it does not change its sign.

Seo and Serizawa [48] measured the changes in surface stress for a (111)-textured Pt thin-film electrode in sulfate solutions with different pH values by using a cantilever bending method. Figure 3.14 shows (a) the cyclic voltammogram, (b) the Δg versus E curve, and (c) the $\frac{\partial g}{\partial E}$ versus E curve obtained at a potential scan rate of 20 mV s^{-1} for the (111)-textured Pt thin-film electrode in pH 2.4, 0.5 M sulfate solution [48]. The $\frac{\partial g}{\partial E}$ versus E curve in Fig. 3.14c is obtained by differentiating Δg with respect to E for the Δg versus E curve in the anodic potential scan of −0.26 to 1.15 V (SHE) in Fig. 3.14b. The value of Δg decreases gradually toward compressive direction with increasing potential in the electric double-layer region where a capacitive charging current flows mainly and it decreases rapidly in the oxide region where oxide formation/reduction takes place. On the other hand, Δg decreases rapidly with decreasing potential in the hydrogen region where hydrogen adsorption/desorption takes place on Pt. The $\frac{\partial g}{\partial E}$ versus E curve in Fig. 3.14c indicates that the sign of $\frac{\partial g}{\partial E}$ or $\zeta_{g,q}$ is plus in the hydrogen region, while it becomes minus in both the electric double-layer and oxide regions. The sign-reversal of $\frac{\partial g}{\partial E}$ or $\zeta_{g,q}$ in the hydrogen region has been also observed for the (111)-textured Pt thin-film electrode in pH 12.3, 0.5 M NaF solution [48]. The sign-reversal of $\frac{\partial g}{\partial E}$ or $\zeta_{g,q}$ in the hydrogen region is consistent with the results obtained with the piezoelectric technique [44, 45]. Moreover, the sign-reversal of $\frac{\partial g}{\partial E}$ or $\zeta_{g,q}$ from minus to plus in the oxide region has been observed by the piezoelectric technique [44, 45]. However, the sign-reversal was not observed in the oxide region by the cantilever bending method [48].

Viswanath et al. [49, 50] measured the changes in surface stress of a consolidated cylindrical nano-porous Pt electrode (particle size: 6 ± 1 nm and mass-specific surface area: $\alpha_m = 25.3$ m^2 g^{-1}) in 0.7 M NaF solution by in situ dilatometry (see Sect. 2.5 of Chap. 2) and found that the sign-reversal of $\zeta_{g,q}$ takes place, depending on the potential region and the surface condition of the nano-porous Pt electrode. In principle, the relative length change $\frac{\Delta l}{l_o}$ of the nano-porous Pt electrode measured by a dilatometer can be converted to the mean value of surface stress change $\langle \Delta g \rangle_A$ as represented by Eq. (2.40) in Sect. 2.5 of Chap. 2: $\langle \Delta g \rangle_A = -\frac{9Y_b}{2\alpha_m \rho}\frac{\Delta l}{l_o}$, where Y_b is the bulk modulus, α_m is the mass-specific surface area, and ρ is the mass density. Figure 3.15 shows the relationship between $\frac{\Delta l}{l_o}$ or $-\Delta g$ and net surface charge ΔQ obtained during cyclic voltammetry in different potential regions between (a) −0.95 and 1.20 V (SHE), (b) −0.30 and 1.10 V (SHE), and (c) 0.30 and 1.20 V (SHE) at a potential scan rate of 1 mV s^{-1} by in situ dilatometry for the nano-porous Pt electrode in 0.7 M NaF solution [50]. In Fig. 3.15a, the arrows from HA and HD represent the progress directions of hydrogen adsorption and desorption, respectively, while the arrows from OA and OD represent the progress directions of oxygen (or OH) adsorption and desorption, respectively. The relationship between $\frac{\Delta l}{l_o}$ or $-\Delta g$ and ΔQ in the HA/HD region has an opposite slope as compared to that in the OA/OD region, i.e., $\zeta_{g,q} = \frac{\Delta g}{\Delta Q} > 0$ in the HA/HD region, while $\zeta_{g,q} = \frac{\Delta g}{\Delta Q} < 0$ in the OA/OD region, from which the sign-reversal of $\zeta_{g,q}$ in the hydrogen region is confirmed.

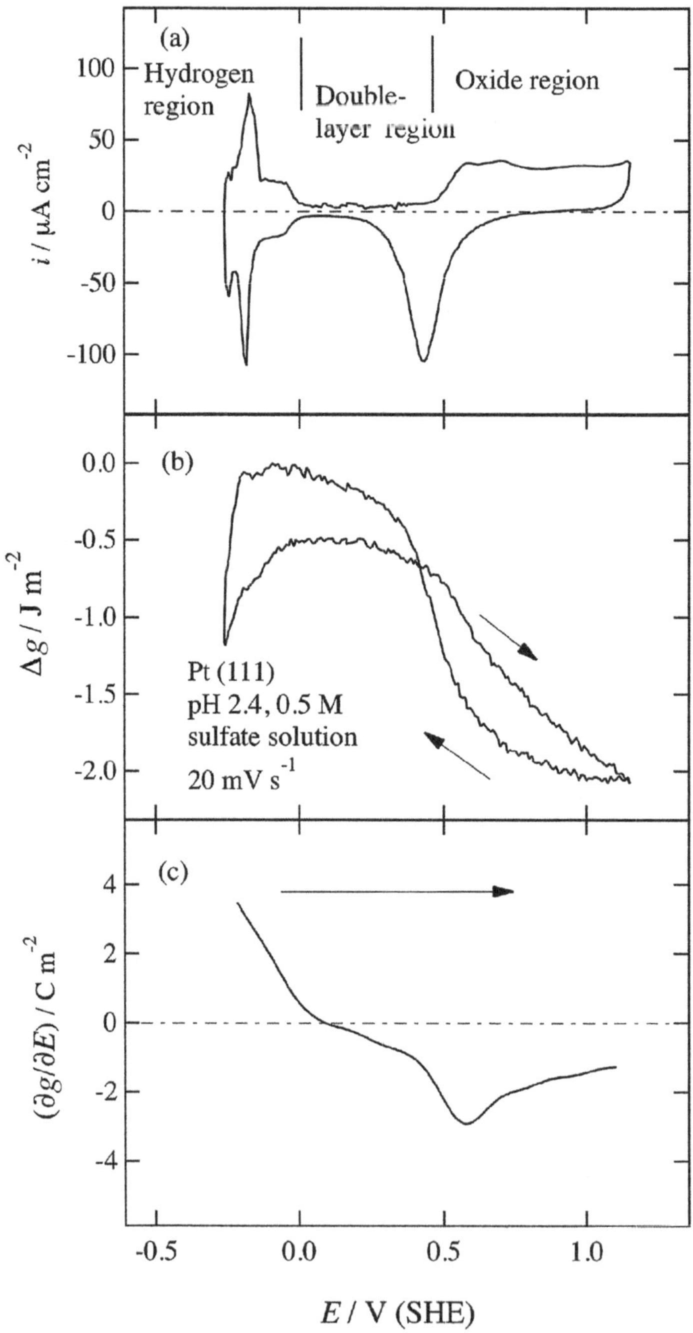

Fig. 3.14 **a** Cyclic voltammogram, **b** the Δg versus E curve, and **c** the $\frac{\partial g}{\partial E}$ versus E curve obtained at a potential scan rate of 20 mV s^{-1} for the (111)-textured Pt thin-film electrode in pH 2.4, 0.5 M sulfate solution [48]. The $\frac{\partial g}{\partial E}$ versus E curve is obtained by differentiating Δg with respect to E for the Δg versus E curve in the anodic potential scan of −0.26 to 1.15 V (SHE) in Fig. 3.14b. Modified from [48] with permission from The Electrochemical Society

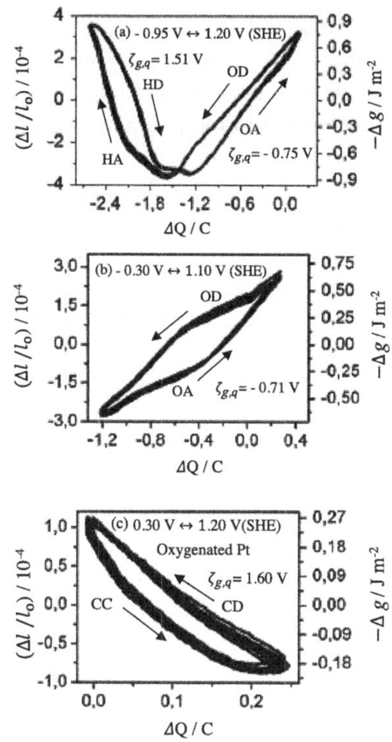

Fig. 3.15 Relationship between $\frac{\Delta l}{l_o}$ or $-\Delta g$ and net surface charge ΔQ obtained during cyclic voltammetry in different potential regions between **a** −0.95 and 1.20 V (SHE), **b** −0.30 and 1.10 V (SHE), and **c** 0.30 and 1.20 V (SHE) at a potential scan rate of 1 mV s^{-1} by in situ dilatometry for the nano-porous Pt electrode in 0.7 M NaF solution [50]. HA: hydrogen adsorption, HD: hydrogen desorption, OA: oxygen adsorption, OD: oxygen desorption, CC: electric double-layer charging, and CD: electric double-layer discharging. Reprinted from [50], Copyright 2007, with permission from Elsevier

The average values of $\zeta_{g,q} = 1.51$ V and $\zeta_{g,q} = -0.75$ V are obtained in the hydrogen and oxygen regions, respectively. The cyclic voltammogram of the nano-porous Pt electrode in pH-neutral solution such as NaF solution exhibits the cathodic shifts of the O-desorption and H-adsorption peaks as well as the anodic shifts of the H-desorption and O-adsorption peaks as compared to the cyclic voltammogram of the planar polycrystalline Pt electrode [50], since the transport kinetics in the pore of Pt limits the electrode processes due to the potential gradient in the pore space. Despite the large hysteresis for the cyclic voltammogram of the nano-porous Pt electrode, the relationship between $\frac{\Delta l}{l_o}$ or $-\Delta g$ and ΔQ in Fig. 3.15a separates sufficiently the two processes of HA/HD and OA/OD. In Fig. 3.15b, although the potential region is limited to the OA/OD process, the average value of $\zeta_{g,q} = -0.71$ V is nearly equal to that for the OA/OD process in Fig. 3.15a.

In order to prepare the oxygenated surface, the cyclic voltammetry of the nano-porous Pt electrode was performed in 0.7 M NaF solution in the potential range of

0.45–1.3 V (SHE), more positive than the reduction peak of surface oxide. The net anodic charge transferred after 12 cycles is about 6 C which corresponds to about 1.5 monolayers of PtO. Afterward, the cyclic voltammetry of the oxygenated nano-porous Pt electrode was performed in the potential range of 0.3–1.20 V (SHE) to measure simultaneously the changes in surface stress by in situ dilatometry as shown in Fig. 3.15c. The capacitive charging (CC) or discharging (CD) current flows mainly for the oxygenated Pt surface in the above potential range. The average value of $\zeta_{g,q} = 1.60$ V is obtained from the slope of the relationship between $\frac{\Delta l}{l_o}$ or $-\Delta g$ and ΔQ in the CC and CD directions in Fig. 3.15c.

In order to prepare the clean (oxide-free) nano-porous Pt electrode as the next step, the cyclic voltammetry was performed in the potential range of −0.35 to 0.1 V (SHE) until the cathodic current of oxide reduction became negligible. Afterward, the cyclic voltammetry of the clean nano-porous Pt electrode was performed in the potential range of 0.0–0.70 V (SHE) to measure simultaneously the changes in surface stress. The capacitive charging (CC) or discharging (CD) current flowed mainly for the clean Pt surface in the above potential range. Although not shown in Fig. 3.15, the average value of $\zeta_{g,q} = -1.06$ V for the clean Pt surface [50] was obtained from the slope of the relationship between $\frac{\Delta l}{l_o}$ or $-\Delta g$ and ΔQ in the CC and CD directions. It is remarked that the sign of $\zeta_{g,q}$ for the oxygenated Pt surface is opposite to the sign of $\zeta_{g,q}$ for the clean Pt surface, which supports the results of the polycrystalline Pt foil electrodes obtained by the piezoelectric detection [44, 45].

The sign-reversal of $\zeta_{g,q}$ has been also observed for a (111)-textured Pt thin-film electrode in 0.1 M $HClO_4$ solution by using the DSA/EIS method [40]. Figure 3.16 shows (a) the real component ζ_r and (b) the phase angle $\psi_{g,q}$ of $\zeta_{g,q}$ as a function of potential for the (111)-textured Pt thin-film electrode, responding to the input signal amplitude of 50 mV at a frequency of 1 Hz. In the dotted curve of Fig. 3.16, the applied dc potential E_{dc} is sequentially stepped by 25 mV, starting from −0.1 to 0.4 V (SSE), then from 0.4 to −0.6 V (SSE), and finally back to −0.1 V (SSE), while in the solid curve, E_{dc} is sequentially stepped by 30 mV, starting from 0.0 to 0.6 V (SSE), then from 0.6 to −0.6 V (SSE), and finally back to 0.0 V (SSE). The value of the imaginary component ζ_i remained around zero over the entire potential region between −0.6 and 0.6 V (SSE) although not shown in Fig. 3.16 [40].

The feature of ζ_r versus E curve in Fig. 3.16a is quite similar to that of the $\frac{\partial g}{\partial E}$ versus E curve in Fig. 3.14c. The potential regions of hydrogen, electric double layer, and oxide described in Fig. 3.16b were determined from the cyclic voltammogram of the (111)-textured Pt thin-film electrode measured in 0.1 M $HClO_4$ solution at a potential scan rate of 50 mV s^{-1} [40]. The positive value of ζ_r in the hydrogen region for the (111)-textured Pt in 0.1 M $HClO_4$ solution is consistent with the results by in situ dilatometry for the nano-porous Pt electrode in 0.7 M NaF solution [50]. Furthermore, the positive value of ζ_r decreases with increasing potential and the sign of ζ_r changes from plus to minus, accompanying the gradual change in $\psi_{g,q}$ from 0° to 180° in the electric double-layer region. The negative value of ζ_r in the electric double-layer region is also consistent with the results by in situ dilatometry for the nano-porous Pt electrode [50].

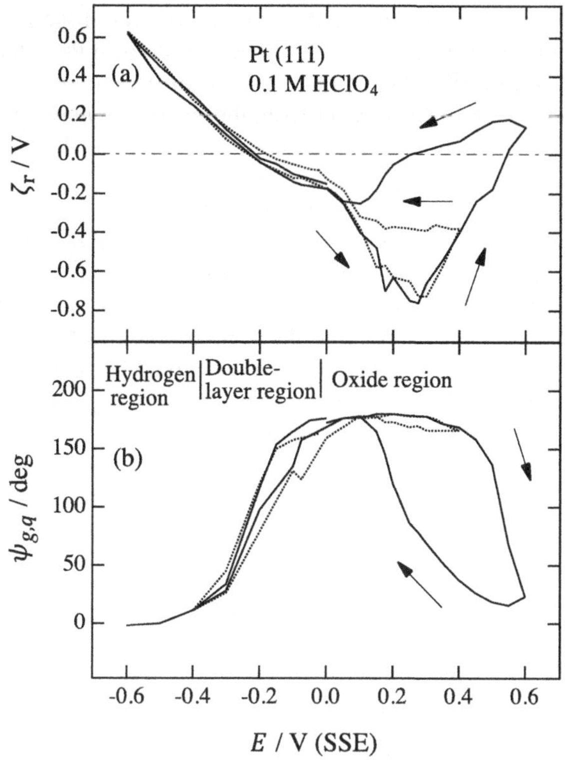

Fig. 3.16 **a** Real component ζ_r and **b** the phase angle $\psi_{g,q}$ of $\zeta_{g,q}$ obtained as a function of potential for the (111)-textured Pt cantilever electrode in 0.1 M $HClO_4$ solution, responding to the input signal amplitude of 50 mV at a frequency of 1 Hz [40]. Reproduced from [40] with permission from The Electrochemical Society

The sign of ζ_r keeps minus in the oxide region during anodic potential scan and ζ_r takes the most negative value (−0.75 V) at about 0.27 V (SSE). In the anodic potential limit of 0.4 V (SSE) for the dotted curve of Fig. 3.16, the hysteresis of ζ_r between sequential anodic and cathodic steps in the oxide region is small, $\psi_{g,q}$ keeping constant around 180°, while in the anodic potential limit of 0.6 V (SSE) for the solid curve of Fig. 3.16, ζ_r exhibits a large hysteresis, accompanying the gradual changes of $\psi_{g,q}$ from 180° to around 0° in the sequential anodic steps up to 0.6 V (SSE) and from around 0° to 180° in the sequential cathodic steps from 0.6 V (SSE). Particularly, the sign of ζ_r changes from minus to plus at about 0.55 V (SSE) in the sequential anodic steps and from plus to minus at about 0.2 V (SSE) in the sequential cathodic steps from 0.6 V (SSE) in response to the gradual changes of $\psi_{g,q}$ between 180° and 0°.

The plus sign of ζ_r with the change of $\psi_{g,q}$ by 180° in the potential range of 0.6 to 0.3 V (SSE) is associated with the formation of surface PtO monolayer [44, 45]. In the low potential range of 0.1 to 0.3 V (SSE) within the oxide region, chemisorbed oxygen (or OH) would reside on the Pt surface. However, if the potential exceeds 0.4 V (SSE), a place-exchange of chemisorbed oxygen with Pt surface atoms takes place to form a surface PtO lattice [44]. The positive value of ζ_r in the potential range from 0.6 to 0.3 V (SSE) is inherent in the surface PtO layer formed with the place-exchange mechanism, which is supported by the in situ dilatometry results of the oxygenated porous Pt electrode in Fig. 3.15c.

The sign-reversal of $\zeta_{g,q}$ has been observed for Pd [51–54] as well as Pt. The positive value of $\zeta_{g,q} = 1.7 \pm 0.3$ V in the hydrogen adsorption/desorption region and the negative value of $\zeta_{g,q} = -1.07 \pm 0.07$ V in the capacitive charging region or $\zeta_{g,q} = -0.86 \pm 0.15$ V in the oxygen adsorption/desorption region were measured by in situ dilatometry for a nano-porous Pd electrode in 0.7 M NaF solution [51, 53]. In addition, $\zeta_{E,\varepsilon} = 1.1$ V in the hydrogen adsorption/desorption region, $\zeta_{E,\varepsilon} = -1.3$ V in the electric double layer or oxygen adsorption/desorption region, and $\zeta_{E,\varepsilon} = 0.5$ V in the PdO formation/reduction region have been reported for a (111)-textured Pd thin-film electrode in 0.01 M H_2SO_4 solution [52, 54]. Particularly, the feature of the $\zeta_{E,\varepsilon}$ versus E curve is quite similar to the results obtained by the piezoelectric detection for the polycrystalline Pd foil electrode in 0.5 M sulfate solutions with pH values of 1.87, 8.8, and 12.8 [46].

3.4.5 *Origin of Sign-Reversal of* $\zeta_{g,q}$

Feibelman [55] made first-principles calculations of the effects of hydrogen and oxygen adsorption on surface stress for a Pt (111) surface to indicate that both hydrogen and oxygen adsorptions reduce tensile stress on Pt (111) terraces despite the fact that hydrogen adsorption reduces the work function of the substrate Pt, while oxygen adsorption increases the work function. The calculated surface stress for hydrogen adsorption on Pt (111) decreases linearly from 6.27 to 1.70 J m^{-2} with increasing hydrogen coverage θ_H from zero to monolayer, i.e., $\Delta g = -4.57$ J m^{-2} for the H (1 × 1)/Pt (111) system. Since the cathodic charge density required for the formation of hydrogen monolayer on Pt (111) surface is $\Delta q_c = -2.4$ C m^{-2}, the positive value of $\zeta_{g,q} = \frac{\Delta g}{\Delta q_c} = 1.9$ V is estimated for the Pt (111) surface in the hydrogen region, which is close to $\zeta_{g,q} = 1.51$ V obtained in the hydrogen region for the nano-porous Pt electrode in 0.7 M NaF solution [50].

The decrease in work function due to hydrogen adsorption means that adsorbed hydrogen atoms are electron-donating adsorbates. If the donated electrons enter the unfilled bonding states of Pt, the bonds in the uppermost surface of Pt would be reinforced, increasing the tensile stress. In contrast to the above speculation, the unfilled 5d-bands of Pt are antibonding, and thereby the donation of electrons into the unfilled antibonding states of Pt should weaken the inter-Pt bonds to reduce the tensile stress [55]. The plus sign of $\zeta_{g,q}$ in the hydrogen region for Pt may result from the electron donation into the unfilled antibonding states of Pt. Pd as well as Pt takes the positive value of $\zeta_{g,q}$ in the hydrogen region. As suggested by Weissmüller [54], the extra electrons donated from hydrogen would contribute to filling the antibonding upper d-band states just above the Fermi level of Pd, and thereby reducing the tensile stress.

Furthermore, Feibelman [55] calculated $\Delta g = -3.2\ \mathrm{J\ m^{-2}}$ for the fcc-O (2 × 2)/Pt (111) system corresponding to the oxygen adsorption of 1/4 monolayer on Pt (111). Since the anodic charge density required for the formation of 1/4 oxygen monolayer on Pt (111) is $\Delta q_{\mathrm{a}} = 1.2\ \mathrm{C\ m^{-2}}$, the negative value of $\zeta_{g,q} = \frac{\Delta g}{\Delta q_{\mathrm{a}}} = -2.7$ V is estimated for the oxygen adsorption on Pt (111) surface [55]. The minus sign of $\zeta_{g,q}$ is consistent with the experimental results obtained in the oxygen adsorption/desorption region by in situ dilatometry for the porous Pt electrode in 0.7 M NaF solution [49, 50] and by DSA/EIS for the (111)-textured Pt thin-film electrode in 0.1 M $HClO_4$ solution [40]. Albina et al. [56] made the first-principles electronic structure calculations based on density functional theory (DFT) with the local density approximation (LDA) to obtain the values of $\zeta_{g,q}$ for transition and noble metals. The value of $\zeta_{g,q}$ for a metal electrode with no surface charge ($q = 0$) corresponding to potential of zero charge in electrolyte can be calculated from the strain dependence of work function in vacuum [56]:

$$\zeta_{g,q} = \left(\frac{\partial g}{\partial q}\right)_{\varepsilon} = \frac{1}{e_{\mathrm{o}}}\left(\frac{\partial \Phi}{\partial \varepsilon}\right)_{q}, \tag{3.26}$$

where Φ is the work function (eV: 1.602×10^{-19} J) and e_{o} is the elementary charge (1.602×10^{-19} C). The values of $\zeta_{g,q} = -1.0$ and -0.98 V were calculated, respectively, for Pt (111) and Pd (111) electrodes, which are consistent with the experimental results obtained in the electric double-layer region by in situ dilatometry for the nano-porous Pt and Pd electrodes in 0.7 M NaF solution [49–51, 53] and DECMA for the (111)-textured Pd electrode in 0.01 M H_2SO_4 solution [52, 54], indicating that the value of $\zeta_{g,q}$ for the metal electrode in the double-layer region, where the adsorption effects are absent or negligible, is equivalent to that for the metal in vacuum.

The minus sign of $\zeta_{g,q}$ in the electric double-layer region may be explained in terms of electrocapillarity effect by Weissmüller's group [37, 50, 56, 57] as follows: (1) The excess electronic charge is localized on the uppermost surface due to large electronic screening effect of fcc transition metal, (2) when the excess electronic charge is negative, i.e., positive excess of electrons ($q < 0$) at potentials more negative than E_{pzc}, the electrostatic center of gravity of the surface Wigner–Seitz cell shifts to promote an outward stretch of the ion cores [58], and (3) insofar as the excess electronic charge does not enter the orbitals responsible for the bonding between the surface atoms, the excess electronic charge is redistributed into the in-plane bonds, which contributes to the increase in tensile surface stress $\Delta g > 0$, thereby leading to the minus sign of $\zeta_{g,q}$. In the case where the excess electronic charge is positive, i.e., negative excess of electrons ($q > 0$) at potentials more positive than E_{pzc}, the electrostatic center of gravity of the surface Wigner–Seitz cell shifts to promote an inward contraction of the ion cores, and the negative excess of electronic charge provides the decrease in tensile stress $\Delta g < 0$, thereby leading to the same minus sign of $\zeta_{g,q}$.

By the way, it seems difficult to explain convincingly the reason why the sign-reversal of $\zeta_{g,q}$ from minus to plus takes place at the transition from oxygen adsorption layer to surface oxide layer (PtO or PdO). Viswanath et al. [50] tried to explain the sign-reversal of $\zeta_{g,q}$ by regarding the surface oxide layer as a semiconductor with weak electronic screening. If the electronic screening on the PtO surface is less effective than on the clean Pt surface, the excess charge would be able to penetrate deeper. In the case where the oxide-covered surface is more negatively charged ($q < 0$), the electrons penetrated in the oxide layer may fill the antibonding valence band states to weaken the Pt–O bonds, expanding the surface, i.e., $\Delta g < 0$, thereby leading to the plus sign of $\zeta_{g,q}$. Conversely, in the case where the oxide-covered surface is more positively charged ($q > 0$), the extraction of electrons from the oxide layer may deplete the antibonding valence band states to strengthen the Pt-O bonds, contracting the surface, i.e., $\Delta g > 0$, thereby leading to the plus sign of $\zeta_{g,q}$.

Apart from $\zeta_{g,q}$, let us discuss briefly the sign-reversal of $\left(\frac{\partial g}{\partial E}\right)_{\varepsilon}$. The Gokhshtein equation (see Eq. (1.119) in Sect. 1.8 of Chap. 1) indicates that the sign-reversal of $\left(\frac{\partial g}{\partial E}\right)_{\varepsilon}$ takes place when the inequality of $-q < \left(\frac{\partial q}{\partial \varepsilon}\right)_E$ is reversed to $-q > \left(\frac{\partial q}{\partial \varepsilon}\right)_E$. It is deduced that the sign-reversal of $\left(\frac{\partial g}{\partial E}\right)_{\varepsilon}$ observed by the piezoelectric detection in the oxide formation/reduction region of the Pt [44, 45] and Pd [46, 47] electrodes results from the significant change of $\left(\frac{\partial q}{\partial \varepsilon}\right)_E$ accompanying the sign-reversal. The change in electronic structure of the Pt or Pd electrode surface due to surface oxide formation and reduction should be inevitably linked to the change of $\left(\frac{\partial q}{\partial \varepsilon}\right)_E$.

References

1. Kolb DM (1993) Surface reconstruction at metal-electrolyte interfaces. In: Lipkowski J, Ross PN (eds) Structure of electrified interfaces. VCH Publishers Inc, New York, pp 65–102
2. Kolb DM (1996) Prog Surf Sci 51:109–173
3. Van Hove MA, Koestner RJ, Stair PC, Bibérian JP, Koesmodel KK, Bartos I, Somorjai GA (1981) Surf Sci 103:189–238
4. Moritz W, Wolf D (1985) Surf Sci 163:L655–L665
5. Möller J, Niehus H, Heiland W (1986) Surf Sci 166:L111–L114
6. Bott M, Hohage M, Michely T, Comsa G (1993) Phys Rev Lett 70:1489–1492
7. Ibach H (2006) Physics of surfaces and interfaces. Springer-Verlag, Berlin
8. Ibach H, Bach CE, Giesen M, Grossmann A (1997) Surf Sci 375:107–119
9. Bach CE, Giesen M, Ibach H (1997) Phys Rev Lett 78:4225–4228
10. Kolb DM, Schneider J (1985) Surf Sci 162:764–775
11. Eberhardt D, Santos E, Schmickler W (1996) J Electroanal Chem 419:23–31
12. Ibach H, Santos E, Schmickler W (2003) Surf Sci 540:504–507
13. Needs RJ, Mansfield M (1989) J Phys: Condens Matter 1:7555–7563
14. Payne MC, Roberts N, Needs RJ, Needels M, Joannopoulos JD (1989) Surf Sci 211/212:1–20
15. Fiorentini V, Methfessel M (1993) Scheffler. Phys Rev Lett 71:1051–1054
16. Haftel MI, Rosen M (2001) Phys Rev B 64:195405-1-7
17. Gao X, Edens GJ, Hamelin A, Weaver MJ (1993) Surf Sci 296:333–35
18. Santos E, Schmickler W (2004) Chem Phys Lett 400:26–29

19. Gao X, Hamelin A, Weaver M (1991) Phys Rev Lett 67:616–622
20. Ibach H (1997) Surf Sci Rep 29:193–263
21. Haiss W, Sass JK (1995) J Electroanal Chem 386:267–270
22. Haiss W, Nichols RJ, Sass JK, Charle KP (1998) J Electroanal Chem 452:199–202
23. Schmickler W, Leiva E (1998) J Electroanal Chem 453:61–67
24. Ibach H (1999) Electrochim Acta 45:575–581
25. Barth JV, Brune H, Ertl G, Behm RJ (1990) Phys Rev B 42:9307–9318
26. Haiss W, Sass JK (1996) Langmuir 12:4311–4313
27. Magnussen OM, Haengeboeck J, Hotlos J, Behm RJ (1992) Faraday Dicuss 94:329–338
28. Shi Z, Wu S, Lipkowski J (1995) Electrochim Acta 40:9–15
29. Shi Z, Lipkowski J (1994) J Electroanal Chem 364:289–294
30. Kramer D (2008) Phys Chem Chem Phys 10:168–177
31. Ocko BM, Watson GM, Wang J (1994) J Phys Chem 98:897–906
32. Ueno K, Seo M (1999) J Electrochem Soc 146:1496–1499
33. Weaver MJ, Gao X (1993) Annu Rev Phys Chem 44:459–494
34. Gokhshtein AY (1976) Surface tension of solids and adsorption. Nauka, Moscow
35. Smetanin M, Viswanath RN, Kramer D, Beckmann D, Koch T, Kilber LA, Kolb DM, Weissmüller J (2008) Langmuir 24:8561–8567
36. Kolb DM, Schneider J (1986) Electrochim Acta 1:929–936
37. Umeno Y, Elsässer C, Meyer B, Gumbsch P, Nothacker M, Weissmüller J, Evers F (2007) Europhys Lett 78:13001-1-5
38. Tabard-Cossa V, Godin M, Burgess IJ, Monga T, Lennox RB, Grtter P (2007) Anal Chem 79:8136–8143
39. Lafouresse MC, Bertocci U, Beauchamp CR, Stafford GR (2012) J Electrochem Soc 159:H816–H822
40. Lafouresse MC, Bertocci U, Beauchamp CR, Stafford GR (2013) J Electrochem Soc 160:H636–H643
41. Sader JE (1998) J Appl Phys 84:64–76
42. Smetanin M, Deng Q, Weissüller J (2011) Phys Chem Chem Phys 13:17313–17322
43. Smetanin M, Kramer D, Mohanan S, Herr U, Weissmüller J (2009) Phys Chem Chem Phys 11:9008–9012
44. Seo M, Makino T, Sato N (1986) J Electrochem Soc 133:1138–1142
45. Jiang XC, Seo M, Sato N (1991) J Electrochem Soc 138:137–140
46. Seo M, Aomi M (1992) J Electrochem Soc 139:1087–1090
47. Seo M, Aomi M (1993) J Electroanal Chem 347:185–194
48. Seo M, Serizawa Y (2003) J Electrochem Soc 150:E472–E476
49. Viswanath RN, Kramer D, Weissmüller J (2005) Langmuir 21:4604–4609
50. Viswanath RN, Kramer D, Weissmüller J (2008) Electrochim Acta 53:2757–2767
51. Weissmüller J, Viswanath RN, Kibler LA, Kolb DM (2011) Phys Chem Chem Phys 13:2114–2117
52. Deng Q, Smetanin M, Weissüller J (2012) 221st ECS Meeting Abstract MA2012-01(23):959
53. Viswanath RN, Weissüller J (2013) Acta Mater 61:6301–6309
54. Weissmüller J (2013) Electrocapillarity of solids and its impact on heterogeneous catalysis. In: Alkire RC, Kolb DM, Kibler LA, Lipkowski J (eds) Advances in electrochemical science and engineering. vol 14. Wiley-VCH, Weinheim
55. Feibelman PJ (1997) Phys Rev B 56:2175–2182
56. Albina I-M, Elsässer C, Weissmüller J, Gumbsch P, Umeno Y (2012) Phys Rev B 85:125118-1-5
57. Umeno Y, Elsässer C, Meyer B, Gumbsch P, Weissmüller J (2008) Europhys Lett 84:13002-1-6
58. Weigend F, Evers F, Weissmüller J (2006) Small 2:1497–1503

Chapter 4
Changes in Surface Stress Associated with Underpotential Deposition and Surface Alloying

Abstract Surface stress varies during underpotential deposition (UPD) of metal atoms on foreign metal electrodes in electrolyte solutions containing the corresponding metallic cations. The definition of UPD, the relationship between UPD and work function, and the equilibrium potential of UPD are explained to understand the UPD phenomena. The typical results of the changes in surface stress induced by the formation and structural changes of various UPD layers on a (111)-textured Au thin-film electrode are exemplified, and the main UPD factors influencing the changes in surface stress are discussed. As a result, the bonding of UPD atom with substrate atom and the lattice mismatch between UPD and substrate atoms are listed as important factors influencing the changes in surface stress. Surface alloying/dealloying processes are often involved in structural changes of UPD layers. The changes in surface stress due to surface alloying are also discussed from the energetics viewpoint of UPD layers.

Keywords Surface stress · Underpotential deposition (UPD) · Work function · Structure of UPD layer · Surface alloying

4.1 Introduction

Underpotential deposition (UPD) of metal atoms (e.g., Pb, Bi, and Cu) on foreign noble metal electrodes such as Au, Pt, and Pd in electrolyte solutions containing the corresponding metallic cations has been extensively investigated since the electrocatalytic activities of the noble metal electrodes for the electrochemical reactions such as oxygen reduction, hydrogen evolution, or oxidation of organic compounds are often enhanced by the UPD treatments of the electrode surfaces [1, 2]. The UPD potential window in which the UPD proceeds is directly related to the difference in work function between the UPD metal and the substrate metal. The equilibrium potential of the UPD depends on the surface coverage of the UPD metal atom and the activity of the corresponding metallic cations in solution.

The changes in surface stress during various UPD processes on the noble metal electrodes can be measured by using a cantilever bending method. In addition, various in situ surface analytical tools with an atomic resolution have been recently developed

M. Seo, *Electro-Chemo-Mechanical Properties of Solid Electrode Surfaces*,
https://doi.org/10.1007/978-981-15-7277-7_4

to reveal the formation processes and structural changes of various UPD layers on noble metal electrodes. The correlation between the changes in surface stress and the structural changes of UPD layers is one of the most interesting topics from the electro-chemo-mechanical viewpoint.

This chapter deals mainly with the changes in surface stress induced by the formation and structural changes of various UPD layers on a (111)-textured Au thin-film electrode. In some UPD systems, co-adsorption of electrolyte anions with UPD atoms stabilizes the UPD layers. We discuss the main UPD factors influencing the changes in surface stress including co-adsorption of electrolyte anions with UPD atoms. Furthermore, surface alloying/dealloying processes are often involved in structural changes of UPD layers. The changes in surface stress due to surface alloying are closely linked to the stability of the UPD layers. We also discuss the relationship between the changes in surface stress due to surface alloying and the stability of the UPD layers from the viewpoint of energetics.

4.2 Underpotential Deposition (UPD)

4.2.1 UPD and Work Function

The term of underpotential deposition (UPD) is applied to the reaction for metallic cations M^{z+} in solution being electrodeposited up to the monolayer level on foreign metal electrode M′ at potentials more positive than the equilibrium potential of M^{z+}/M electrode [1, 3]. Kolb et al. [1, 3] found empirically that the UPD potential window ΔE_{UPD} is proportional to the difference in work function between electrodeposited metal M and substrate metal M′:

$$\Delta E_{\mathrm{UPD}} \approx 0.5(\Phi_{\mathrm{M'}} - \Phi_{\mathrm{M}}), \tag{4.1}$$

where Φ_{M} and $\Phi_{\mathrm{M'}}$ are the work functions of electrodeposited metal M and of substrate metal M′, respectively. Equation (4.1) holds approximately on many UPD systems in which noble metals such as Pt, Au, and Ag are used as substrate metals. The UPD phenomena are originated from the excess binding energy of metal atoms M on substrate metal M′ with respect to metal atoms M on bulk metal M.

As shown schematically in Fig. 4.1, Kolb et al. [1, 3] determined experimentally ΔE_{UPD} as the difference between two characteristic potentials of $E^{a}_{\mathrm{M,b}}$ and $E^{a}_{\mathrm{M,s}}$ in potentiodynamic polarization curve of the substrate metal M′, measured in solution containing metallic cations M^{z+} by anodic potential scan from a potential more negative than the equilibrium potential of M^{z+}/M electrode:

$$\Delta E_{\mathrm{UPD}} = E^{a}_{\mathrm{M,s}} - E^{a}_{\mathrm{M,b}}, \tag{4.2}$$

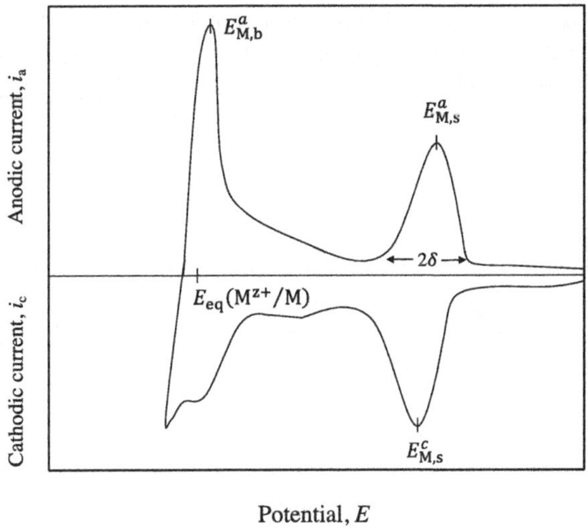

Fig. 4.1 Potential window ΔE_{UPD} of UPD taken as the difference between two characteristic potentials $E^a_{M,b}$, and $E^a_{M,s}$ in potentiodynamic polarization curve of the substrate metal M′, measured in solution containing metallic cations M^{z+} by anodic potential scan from a potential more negative than the equilibrium potential of M^{z+}/M electrode

where $E^a_{M,b}$ and $E^a_{M,s}$ are the potentials at anodic stripping current peaks of bulk-deposited and electrodeposited M, respectively. If the anodic stripping current peak of electrodeposited M is nearly symmetrical in shape, the surface coverage θ_M of M at $E^a_{M,s}$ would be close to 0.5 since the peak area corresponds almost to the monolayer of electrodeposited M. According to Trasatti [4, 5], for the electrodeposition of a metal atom M on substrate metal M′, the deposition energy $W(M' - M)$ at $\theta_M \to 0$ can be written as follows:

$$W\big(M' - M\big) = -I_M + \Phi_{M'} + \frac{e^2}{4x}, \tag{4.3}$$

where I_M is the ionization energy of metal atom M and the third term in the right-hand side of Eq. (4.3) is the image energy (x: distance at the image charge plane from the surface of metal M′ in the jellium model [6]). For the deposition of metal atom M on substrate metal M, the deposition energy $W(M - M)$ at $\theta_M \to 0$ can be also written by

$$W(M - M) = -I_M + \Phi_M + \frac{e^2}{4x}. \tag{4.4}$$

The difference between Eqs. (4.3) and (4.4) is given by

$$W\big(M' - M\big) - W(M - M) = \Phi_{M'} - \Phi_M = e\Delta E_{UPD}(\theta_M \to 0), \tag{4.5}$$

where $\Delta E_{\mathrm{UPD}}(\theta_{\mathrm{M}} \rightarrow 0)$ is the UPD potential window at $\theta_{\mathrm{M}} \rightarrow 0$ which is different from ΔE_{UPD} at $\theta_{\mathrm{M}} \approx 0.5$ in Eq. (4.1) or Eq. (4.2). The value of $\Delta E_{\mathrm{UPD}}(\theta_{\mathrm{M}} \rightarrow 0)$ is equal to that obtained by adding half width δ of the peak at $E^{a}_{\mathrm{M,s}}$ to $\Delta E_{\mathrm{UPD}}(\theta_{\mathrm{M}} \approx 0.5)$ in Fig. 4.1:

$$\Delta E_{\mathrm{UPD}}(\theta_{\mathrm{M}} \rightarrow 0) = \Delta E_{\mathrm{UPD}}(\theta_{\mathrm{M}} \approx 0.5) + \delta. \tag{4.6}$$

Trasatti [5] showed that the linear relationship with unity slope holds between $\Delta E_{\mathrm{UPD}}(\theta_{\mathrm{M}} \rightarrow 0)$ obtained experimentally by Eq. (4.6) and $\Phi_{\mathrm{M'}} - \Phi_{\mathrm{M}}$ for various UPD systems, proving the validity of Eq. (4.5). Equation (4.5) suggests that the surface bond between M and M′ is almost ionic at $\theta_{\mathrm{M}} \rightarrow 0$, which is in similar manner to the initial stage of alkali metal adsorption on jellium metal surface from the gas phase [7].

Figure 4.2 shows schematically the change in work function Φ as a function of surface coverage θ in the case of adsorption of alkali metal atoms such as Na or K on jellium metal surface from the gas phase. The linear decrease in Φ with steep slope at $\theta \approx 0$ results from the formation of dipole due to ionic adsorption of alkali metal. The upward deviation from the linearity with increasing θ is caused by the decrease in dipole moment due to screening with free electrons exuded from jellium metal [7]. The value of Φ increases gradually after taking a minimum in response to the transition from ionic to metallic bond due to the formation of electron energy band in adsorption layer [7]. The coefficient of 0.5 in Eq. (4.1) may come from the employment of anodic stripping peak at $E^{a}_{\mathrm{M,s}}$ ($\theta_{\mathrm{M}} \approx 0.5$) for the determination of ΔE_{UPD} since the surface bond at $E^{a}_{\mathrm{M,s}}$ is not simply ionic but intermediate between

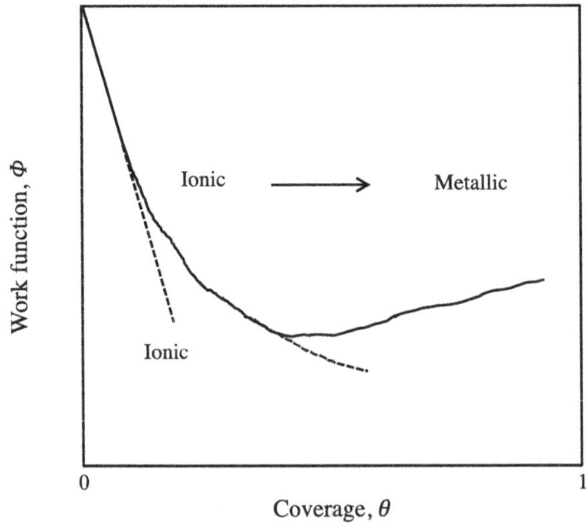

Fig. 4.2 Schematic changes in work function Φ as a function of surface coverage θ in the case of adsorption of alkali metal atoms such as Na or K on jellium metal surface from the gas phase

ionic and metallic as suggested by Trasatti [5]. However, further study on the changes in nature of surface M-M′ bond as a function of θ is needed for better understanding of UPD phenomena.

4.2.2 Equilibrium Potential of UPD and Adsorption Isotherm

At equilibrium, M-UPD on M′ may be described by the following complete discharging reaction:

$$M^{z+} + ze^{-}(M') = M_{ad}(M'). \tag{4.7}$$

The equilibrium potential $E_{sml}(\theta_M)$ for the formation of M sub-monolayer can be defined as

$$E_{sml}(\theta_M) = E^{o}_{sml} + \frac{RT}{zF}\ln\frac{a_{M^{z+}}}{a_{sml}}, \tag{4.8}$$

where E^{o}_{sml} is the standard equilibrium potential for the formation of M sub-monolayer, $a_{M^{z+}}$ is the activity of M^{z+} in solution, and a_{sml} is the activity of M sub-monolayer. Furthermore, the equilibrium potential E_{eq} for the formation of bulk-deposited M layer ($M^{z+} + ze^{-}(M) = M(M)$) is given by

$$E_{eq} = E^{o}_{eq} + \frac{RT}{zF}\ln a_{M^{z+}}. \tag{4.9}$$

The subtraction of Eq. (4.9) from Eq. (4.8) provides

$$E_{sml}(\theta_M) - E_{eq} = E^{o}_{sml} - E^{o}_{eq} - \frac{RT}{zF}\ln a_{sml}. \tag{4.10}$$

Equation (4.10) is further rearranged to

$$E_{sml}(\theta_M) = E^{o}_{sml} - \frac{RT}{zF}\ln a_{sml} + \frac{RT}{zF}\ln a_{M^{z+}}. \tag{4.11}$$

If the M-UPD on M′ obeys the Langmuir type of adsorption isotherm [8], a_{sml} is given by

$$a_{sml} = \frac{\theta_M}{1-\theta_M}. \tag{4.12}$$

Consequently, $E_{\text{sml}}(\theta_{\text{M}})$ can be expressed as a function of θ_{M}:

$$E_{\text{sml}}(\theta_{\text{M}}) = E^{\text{o}}_{\text{sml}} - \frac{RT}{zF} \ln \frac{\theta_{\text{M}}}{1 - \theta_{\text{M}}} + \frac{RT}{zF} \ln a_{\text{M}^{z+}}, \tag{4.13}$$

where $E^{\text{o}}_{\text{sml}}$ is taken as standard at $\theta_{\text{M}} = 0.5$ and $a_{\text{M}^{z+}} = 1$ since $a_{\text{sml}} = 1$ is equivalent to $\theta_{\text{M}} = 0.5$ in Eq. (4.12).

On the other hand, if the M-UPD on M′ obeys the Frumkin type of adsorption isotherm [9], a_{sml} is given by

$$a_{\text{sml}} = \frac{\theta_{\text{M}}}{1 - \theta_{\text{M}}} \exp(-2\beta\theta_{\text{M}}), \tag{4.14}$$

where β is a parameter of the interaction between electrodeposited M atoms. In the case of the repulsive interaction, β takes a positive value, while in the case of the attractive interaction, β takes a negative value. For the Frumkin type of adsorption isotherm, $E_{\text{sml}}(\theta_{\text{M}})$ is formulated by

$$E_{\text{sml}}(\theta_{\text{M}}) = E^{\text{o}}_{\text{sml}} - \frac{RT}{zF} \ln \frac{\theta_{\text{M}}}{1 - \theta_{\text{M}}} + \frac{RT}{zF} \beta\theta_{\text{M}} + \frac{RT}{zF} \ln a_{\text{M}^{z+}}. \tag{4.15}$$

If the M-UPD on M′ obeys the Langmuir type of adsorption isotherm, the UPD potential window $\Delta E_{\text{UPD}}(\theta_{\text{M}} = 0.5)$ defined by Kolb et al. [1, 3] may be given by

$$\Delta E_{\text{UPD}}(\theta_{\text{M}} = 0.5) = E_{\text{eq}} - E_{\text{sml}}(\theta_{\text{M}} = 0.5) = E^{\text{o}}_{\text{eq}} - E^{\text{o}}_{\text{sml}}. \tag{4.16}$$

Equation (4.16) means that $\Delta E_{\text{UPD}}(\theta_{\text{M}} = 0.5)$ is independent of $a_{\text{M}^{z+}}$ in solution since E_{eq} and $E_{\text{sml}}(\theta_{\text{M}} = 0.5)$ exhibit the same $a_{\text{M}^{z+}}$ dependence as represented by Eqs. (4.9) and (4.13).

4.3 Changes in Surface Stress during UPD

In this section, we explain the typical results of the changes in surface stress induced by the formation and structural change of UPD layer for Pb-UPD [10–], Bi-UPD [14, 16], Cu-UPD [17–19], and Pd-UPD [20] on a (111)-textured Au thin-film electrode and discuss the main UPD factors influencing the changes in surface stress from the electr15o-chemo-mechanical viewpoint.

4.3.1 Pb-UPD on Au (111)

Figure 4.3 shows (a) the cyclic voltammogram and (b) the changes in surface stress versus potential (Δg vs. E) curve measured at a potential scan rate of 5 mV s^{-1} for the (111)-textured Au thin-film (with a thickness of 220 nm) electrode in pH 3.0, 0.5 M $NaClO_4$ solution containing 10^{-4} M $Pb(ClO_4)_2$ [12]. In Fig. 4.3, the anodic limit potential is set at 0.80 V (SHE) to avoid the onset of the surface oxygenation reaction on Au, while the cathodic limit potential is chosen at −0.24 V (SHE), corresponding to the equilibrium potential of 10^{-4} M Pb^{2+}/Pb electrode. The changes in surface stress Δg are referred to zero at 0.80 V (SHE).

The Δg vs. E curve (dotted line) measured in pH 3.0, 0.5 M $NaClO_4$ solution without 10^{-4} M $Pb(ClO_4)_2$ is also shown in Fig. 4.3b for comparison. The value

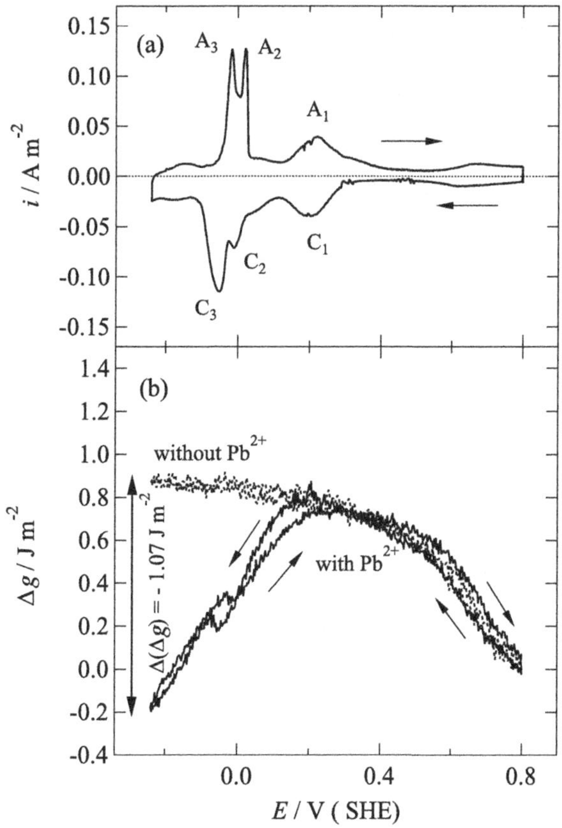

Fig. 4.3 **a** Cyclic voltammogram and **b** the changes in surface stress versus potential (Δg vs. E) curve measured at a potential scan rate of 5 mV s^{-1} for the (111)-textured Au thin-film (with a thickness of 220 nm) electrode in pH 3.0, 0.5 M $NaClO_4$ solution containing 10^{-4} M $Pb(ClO_4)_2$ [12]. The Δg vs. E curve (dotted line) measured in pH 3.0, 0.5 M $NaClO_4$ solution without 10^{-4} M $Pb(ClO_4)_2$ is also shown in Fig. 4.3b for comparison. Reproduced from [12] with permission from The Electrochemical Society

of Δg in the absence of Pb^{2+} varies monotonously with potential in both cathodic and anodic potential scans. It is known that the potential of zero charge E_{pzc} for the unreconstructed Au (111) electrode in perchloric acid is 0.48 V (SHE) [21]. Therefore, the surface tension γ of the Au (111) electrode should take a maximum at 0.48 V (SHE). On the other hand, the Δg vs. E curve in the absence of Pb^{2+} does not take a maximum in the potential region between −0.24 V and 0.80 V (SHE), which is consistent with that reported by Stafford and Bertocci [13]. The Δg vs. E curve in the presence Pb^{2+} has maxima at about 0.2 V and humps at about −0.05 V (SHE) in both cathodic and anodic potential scans. These maxima and humps correspond to the C_1/A_1 current peaks and the C_2/A_2 and C_3/A_3 current peaks, respectively, in the cyclic voltammogram (Fig. 4.3a), which are associated with electrodeposition (UPD) of Pb^{2+} and with anodic stripping of electrodeposited Pb. The Δg vs. E curve in the presence of Pb^{2+} indicates that Δg decreases toward compressive direction with the progress in Pb-UPD at potentials more negative than 0.2 V (SHE). The humps at about - 0.05 V (SHE) result from the structural changes of the Pb-UPD layer as discussed later. The difference $\Delta(\Delta g)$ between the changes in surface stress in the solutions without and with Pb^{2+} attains to -1.07 J m^{-2} at a cathodic limit potential of −0.24 V (SHE).

Figure 4.4 shows the relationship between changes in surface stress Δg and cathodic charge density Δq_c derived from the results of Fig. 4.3 [12]. The value of Δg is taken in the cathodic potential scan from 0.80 V (SHE), while the value

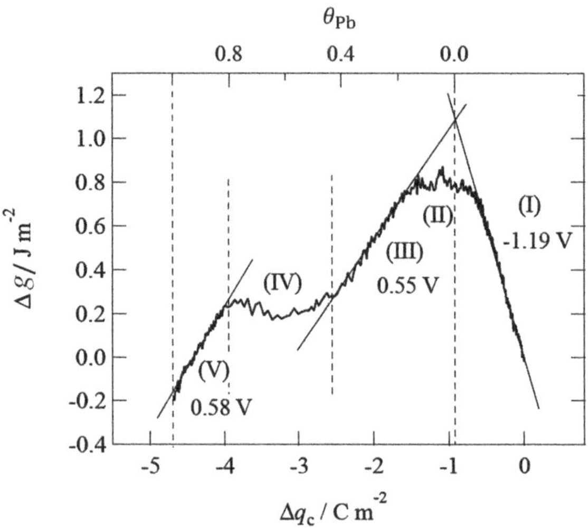

Fig. 4.4 Relationship between changes in surface stress Δg and cathodic charge density Δq_c derived from the results of Fig. 4.3 [12]. The value of Δg is taken in the cathodic potential scan from 0.80 V (SHE), while the value of Δq_c is obtained from the integration of the corresponding cyclic voltammogram, referred to zero at 0.80 V (SHE). The surface coverage θ_{Pb} of Pb on the top abscissa of Fig. 4.4 is calculated from the cathodic charge density by regarding as $\theta_{Pb} = 1.0$ at − 0.24 V (SHE). Reproduced from [12] with permission from The Electrochemical Society

of Δq_c is obtained from the integration of the corresponding cyclic voltammogram, referred to zero at 0.80 V (SHE). In Fig. 4.4, the relationship between Δg and Δq_c consists of three linear (I), (III), and (V), and two plateau (II) and (IV) domains. In the linear (I) domain with a slope of -1.19 V, no UPD of Pb but adsorption of perchlorate ions takes place on the Au (111) electrode. As explained in Sect. 3.4 of Chap. 3, the surface stress–surface charge density coefficient $\zeta_{g,q} = \frac{\partial g}{\partial q}$ takes a minus sign for the Au (111) electrode in the electric double-layer region [22]. The negative slope of $\zeta_{g,q} = \frac{\Delta g}{\Delta q_c} = -1.19\,\text{V}$ in the linear (I) domain is consistent with the minus sign of $\zeta_{g,q}$ for the Au (111) electrode in the electric double-layer region. The plateau (II) domain where surface stress takes a maximum corresponds to the transition from the anion adsorption phase to the Pb-UPD phase. In the linear (III) domain with a slope of 0.55 V, the Pb-UPD proceeds on the Au (111) electrode. The plateau (IV) domain where the hump of surface stress appears corresponds to a structural change of the UPD layer. In the linear (V) domain with a slope of 0.58 V, the Pb-UPD still proceeds until reaching to the UPD monolayer. The positive slopes of 0.55 V in the linear (III) domain and of 0.58 V in the linear (V) domain are characteristic of the Pb-UPD on the Au (111) electrode.

The intersection of two extrapolated lines in the linear (I) and (III) domains at $\Delta q_c = -0.94\,\text{C m}^{-2}$ may be regarded as the onset of the Pb-UPD. The cathodic charge density at the cathodic limit potential of –0.24 V (SHE) is $\Delta q_c = -4.66\,\text{C m}^{-2}$. If the Pb-UPD monolayer is formed at –0.24 V (SHE), the cathodic charge density required for the monolayer formation is $\Delta(\Delta q_c)_m = -3.72\,\text{C m}^{-2}$. The atomic density of an atomically flat Au (111) surface without reconstruction is 1.38×10^{19} atoms m^{-2}. Assuming the complete discharging of Pb^{2+} (i.e., $z = 2$), and the hexagonal close-packed (hcp) Pb monolayer, and taking the difference in radius between Pb ($r = 0.175\,\text{nm}$) and Au ($r = 0.144\,\text{nm}$) atoms into account [12], the cathodic charge density required for the formation of the Pb-UPD monolayer on an atomically flat Au (111) surface is $\Delta(\Delta q_c)_m = -3.02\,\text{C m}^{-2}$. If the surface roughness of the (111)-textured Au film electrode is $S_r = 1.2$, the experimental value of $\Delta(\Delta q_c)_m$ in Fig. 4.4 is close to that theoretically derived.

The surface coverage θ_{Pb} of Pb on the top abscissa of Fig. 4.4 is calculated from the cathodic charge density by regarding as $\theta_{Pb} = 1.0$ at –0.24 V (SHE). The plateau (IV) domain with the hump of surface stress associated with the structural change of the Pb-UPD layer corresponds to $\theta_{Pb} = 0.43 \sim 0.80$. Stafford and Bertocci [13] also measured the changes in surface stress as a function of θ_{Pb} on a (111)-textured Au thin-film electrode in 0.1 M $HClO_4$ solution with 10^{-2} M $Pb(ClO_4)_2$ and observed the same plateau (IV) domain at $\theta_{Pb} = 0.56 \sim 0.86$. According to Stafford and Bertocci [13], the values of θ_{Pb} determined by using an electrochemical quartz microbalance (EQCM) in separated experiments were in good agreement with the coulometric results. The average value of θ_{Pb} at the plateau (IV) domain deviates each other by $\Delta\theta_{Pb} \approx 0.1$, which may result from the difference in ionic strength or activity of Pb^{2+} between the electrolytes used for experiments [12, 13].

According to surface X-ray scattering (SXS) study of the Pb-UPD on Au (111) electrode by Toney et al. [23], the Pb-UPD monolayer with compressive and incommensurate hcp structure is rotated from the gold substrate $[01\bar{1}]$ direction by $R = 2.5°$ at potentials lower than −0.11 V (SHE), but not rotated (i.e., $R = 0°$) at potentials higher than −0.08 V (SHE). The humps of surface stress in Fig. 4.3 were initially ascribed to the relaxation of compressive surface stress caused by the rotation of the incommensurate hcp structure from $R = 0°$ to 2.5° [10–12]. Nevertheless, Stafford and Bertocci [13] claimed that the stress hump is not associated with the rotation of the incommensurate hcp monolayer but with the coalescence of the hcp Pb islands since the potential at which the rotation of the incommensurate hcp monolayer takes place is more negative by about 0.07 V than the potential at which the surface stress hump is observed. The more detailed study by Shin et al. [15] implied that the stress hump is associated with the phase transformation between the incommensurate hcp monolayer and surface alloy of $\left(\sqrt{3} \times \sqrt{3}\right)R30°$. Scanning tunnel microscopic (STM) observation during Pb-UPD on Au (111) [24] indicated that the hcp Pb islands are slowly transformed to the alloy structure of $\left(\sqrt{3} \times \sqrt{3}\right)R30°$ only at a low coverage of Pb, $\theta_{Pb} \approx 0.25$. The transformation between the hcp island and surface alloy structure has been also observed during the Pb-UPD or Tl-UPD on Ag (111) [25].

The alloy structure is usually more thermodynamically stable than the hcp structure at a low Pb coverage [15]. The phase transformation provides an additional shift of surface stress toward compressive direction since Pb atoms with large radius compared to Au atoms are embedded into the Au lattice. The alloy structure develops accompanying the changes in surface stress toward compressive direction in the cathodic potential scan until the Au (111) electrode surface is fully covered with the alloy structure. The full coverage of the surface alloy $\left(\sqrt{3} \times \sqrt{3}\right)R30°$ on a smooth Au (111) surface corresponds theoretically to $\theta_{Pb} = 0.333$. In contrast, in the case where θ_{Pb} exceeds the full coverage of surface alloy, the hcp Pb overlayer would become more stable than the alloy structure because the further embedding of Pb atoms in the alloy phase needs a positive interface energy [15]. Consequently, when the coverage of surface alloy exceeds $\theta_{Pb} = 0.333$, the dealloying process corresponding to the transformation from the alloy structure to the hcp Pb overlayer may occur spontaneously accompanying the relaxation of surface stress toward tensile direction. The surface alloying and dealloying process during the growth of the Pb-UPD layer on Cu (111) has been confirmed by the STM observation [26] in which the full coverage of alloy structure $\left(\sqrt{3} \times \sqrt{3}\right)R30°$ is about $\theta_{Pb} = 0.4$ and the alloy structure is transformed to the hcp Pb overlayer at $\theta_{Pb} > 0.4$. As proposed by Shin et al. [15], it is the most reliable that the stress hump observed during the growth of the Pb-UPD layer on the Au (111) electrode is caused by the transformation between the surface alloy and the incommensurate hcp Pb monolayer.

Figure 4.5 shows the nearest-neighbor distance a_{nn} of the hcp Pb monolayer on Au (111) electrode obtained as a function of potential from the SXS measurements by Toney et al. [23]. The value of $a_{nn} = 0.35$ nm for bulk Pb is taken at −0.05 V (SHE) which is more positive by 0.19 V than the equilibrium potential (−0.24 V) for

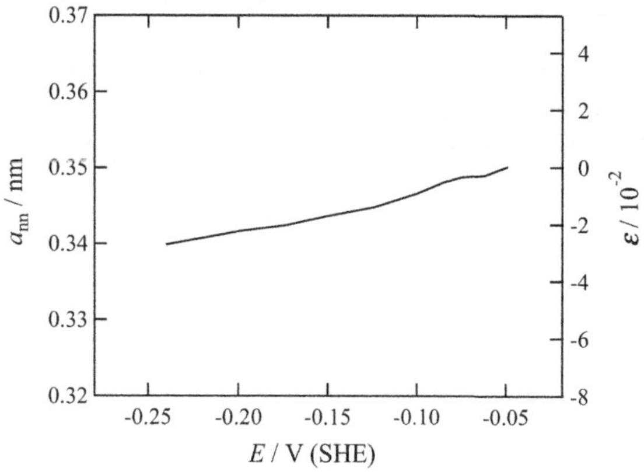

Fig. 4.5 Nearest-neighbor distance a_{nn} of the hcp Pb monolayer on Au (111) electrode obtained as a function of potential from the SXS measurements by Toney et al. [23]. The elastic strain of the hcp Pb monolayer ε on the ordinate in the right-hand side of Fig. 4.5 is converted from a_{nn} by assuming $\varepsilon = 0$ at -0.05 V (SHE). Modified with the permission from [23], Copyright 1995, American Chemical Society

the bulk deposition of Pb. The value of a_{nn} decreases with decreasing potential from -0.05 V (SHE), indicating that the hcp Pb-UPD monolayer is subjected to electro-compression. The elastic strain ε of the hcp Pb-UPD monolayer on the ordinate in the right-hand side of Fig. 4.5 is converted from a_{nn} by assuming $\varepsilon = 0$ at -0.05 V (SHE). Friesen et al. [11] and Stafford and Bertocci [13] converted the surface stress versus potential data into the surface stress versus elastic strain data and found the almost linear relationship between Δg and ε in the electro-compressive region. In Fig. 4.6, the values of Δg measured as a function of potential in the anodic potential scan from -0.24 V (SHE) in Fig. 4.3b are plotted versus ε in the same potential range in Fig. 4.5, indicating the linear relationship between Δg and ε with an average slope of 16.7 J m^{-2} [14]. The average slope of 16.7 J m^{-2} coincides with that (16.8 J m^{-2}) obtained by Stafford and Bertocci [13].

If the hcp Pb-UPD monolayer is regarded as a free-standing elastic film, the relationship between Δg and ε is represented by

$$\Delta g = Y_{(111)}\varepsilon d_{(111)}, \tag{4.17}$$

where $Y_{(111)}$ is the biaxial modulus of the hcp film and $d_{(111)}$ is the film thickness. The value of $Y_{\mathrm{Pb}(111)}d_{(111)} = 16.7$ J m^{-2} for the hcp Pb-UPD monolayer on Au (111) electrode is obtained from the slope of the linear relationship in Fig. 4.6. On the other hand, the value of $Y_{\mathrm{Pb}(111)}d_{(111)}$ for bulk Pb (111) can be calculated since the elastic compliances S_{ij} of bulk Pb are known. Young's modulus $E_{(111)}$, Poisson's ratio $\nu_{(111)}$, and biaxial modulus $Y_{(111)}$ for bulk Pb (111) are derived from the values of S_{ij} [27] as follows:

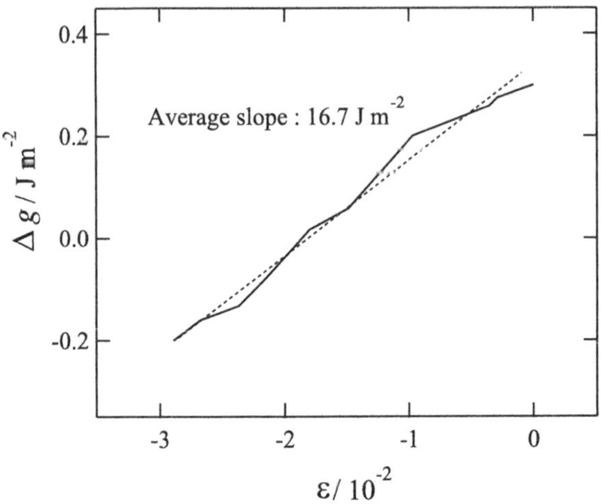

Fig. 4.6 Relationship between Δg and ε in the potential region between −0.24 and −0.05 V (SHE) for the hcp Pb monolayer on Au (111) electrode [14]. The value of Δg is measured as a function of potential in the anodic potential scan from −0.24 V (SHE) in Fig. 4.3b, while the value of ε is taken from Fig. 4.5. Reproduced from [14] with permission from The Electrochemical Society

$$E_{(111)} = \frac{4}{(2S_{11} + 2S_{12} + S_{44})}, \tag{4.18}$$

$$\nu_{(111)} = -E_{(111)} \times \frac{(2S_{11} + 10S_{12} - S_{44})}{12}, \tag{4.19}$$

and

$$Y_{(111)} = \frac{E_{(111)}}{\left(1 - \nu_{(111)}\right)}, \tag{4.20}$$

where $S_{11} = 9.37 \times 10^{-11}\,\text{Pa}^{-1}$, $S_{12} = -4.3 \times 10^{-11}\,\text{Pa}^{-1}$, and $S_{44} = 6.8 \times 10^{-11}\,\text{Pa}^{-1}$ are known for bulk Pb [28]. As a result, $E_{\text{Pb}(111)} = 23.6\,\text{GPa}$, $\nu_{(111)} = 0.61$, and $Y_{\text{Pb}(111)} = 60.5\,\text{GPa}$ are calculated for bulk Pb. The value of $d_{(111)} = 0.2858\,\text{nm}$ is used for bulk Pb since the lattice constant of Pb is 0.495 nm. The value of $Y_{\text{Pb}(111)}d_{(111)} = 17.3\,\text{J m}^{-2}$ is eventually obtained for bulk Pb, and it is close to that (16.7 J m^{-2}) for the hcp Pb-UPD monolayer on Au (111) electrode, which is also in good agreement with the results obtained by Stafford and Bertocci [13].

The value of $Y_{\text{Au}(111)}d_{(111)} = 44.4\,\text{J m}^{-2}$ for the Au (1 × 1) monolayer is calculated as well by using $S_{11} = 2.34 \times 10^{-11}\,\text{Pa}^{-1}$, $S_{12} = -1.07 \times 10^{-11}\,\text{Pa}^{-1}$, $S_{44} = 2.38 \times 10^{-11}\,\text{Pa}^{-1}$, and $d_{(111)} = 0.235\,\text{nm}$ for bulk Au (111), which is about 2.6 times as much as that (16.7 J m^{-2}) for the hcp Pb-UPD monolayer on the Au (111) electrode. It is surprising that the value of $Y_{\text{Pb}(111)}d_{(111)}$ for the hcp Pb-UPD monolayer on the Au (111) electrode is nearly equal to that for the same layer in the bulk Pb, although the reason is not clear. Nevertheless, it has been reported that the elastic constants of even ultra-thin films with lattice mismatch grown on substrates are close to the bulk values [29]. For the Tl-UPD on the (111)-textured Au electrode as well as the Pb-UPD, the hump of surface stress has been observed at the transition potential between surface alloying and dealloying by long-term potentiostatic pulse experiments [30].

4.3.2 *Bi-UPD on Au (111)*

Figure 4.7 shows (a) the cyclic voltammograms and (b) the Δg vs. E curves measured at a potential scan rate of 5 mV s^{-1} for the (111)-textured Au thin-film (with a

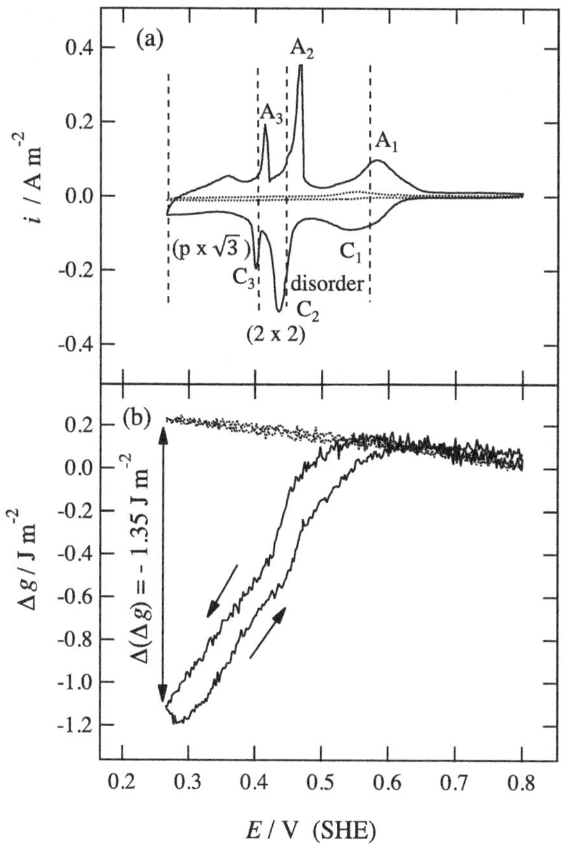

Fig. 4.7 **a** Cyclic voltammograms and **b** the Δg vs. E curves measured at a potential scan rate of 5 mV s^{-1} for the (111)-textured Au thin-film (with a thickness of 220 nm) electrode in 0.1 M $HClO_4$ solutions with and without 2.5×10^{-3} M Bi_2O_3 [14]. The solid and dotted curves represent the results in the solutions with and without Bi^{3+}, respectively. Reproduced from [14] with permission from The Electrochemical Society

thickness of 220 nm) electrode in 0.1 M $HClO_4$ solutions with and without 2.5 × 10^{-3} M Bi_2O_3 [14]. The solid and dotted curves represent the results in the solutions with and without Bi^{3+}, respectively. The cathodic limit potential of 0.26 V (SHE) in Fig. 4.7 corresponds to the equilibrium potential of 5.0 × 10^{-3} M Bi^{3+}/Bi electrode. The anodic limit potential of 0.80 V (SHE) is chosen to avoid the onset of surface oxygenation reaction on Au. Three pairs of cathodic and anodic current peaks (C_1/A_1, C_2/A_2, and C_3/A_3) in the presence of Bi^{3+} are characteristic of the Bi-UPD on the Au (111) electrode. These current peaks are associated with the structural changes of the Bi-UPD layer on the Au (111) electrode.

The structural changes of the Bi-UPD layer on Au (111) in 0.1 M $HClO_4$ solution containing 10^{-3} M Bi^{3+} have been investigated by SXS [31]. The structure of the Bi-UPD layer changes from disorder to (2 × 2) and then to $\left(p \times \sqrt{3}\right)$ – 2Bi with decreasing potential. The stable region of each structure is represented by the dashed vertical lines in Fig. 4.7a. The above structures of the Bi-UPD layer on the Au (111) electrode have been also confirmed by atomic force microscope (AFM) [32] and STM [33, 34]. Furthermore, no surface alloy formation has been reported for the Bi-UPD on the Au (111) electrode [31–34]. The (2 × 2) structure is commensurate with the underlying Au (111) surface, while the $\left(p \times \sqrt{3}\right)$ – 2Bi structure is uniaxially incommensurate with the underlying Au (111) surface. The latter structure is subjected to electro-compression in the incommensurate direction when θ_{Bi} changes from 0.61 to 0.67 with decreasing potential [31]. The pseudomorphic growth of the Bi-UPD layer such as the hcp Pb-UPD layer on the closed-packed Au (111) surface seems difficult since Bi has a rhombohedral crystal structure.

In Fig. 4.7b, the value of Δg is referred to zero at the anodic limit potential of 0.8 V (SHE). The surface stress levels of the Au (111) electrode in the solutions with and without Bi^{3+} are the same at 0.80 V since the Bi-UPD does not proceed at 0.80 V (SHE). As the potential is scanned from 0.80 V (SHE) to the cathodic direction, Δg for the Au (111) electrode in the absence of Bi^{3+} varies monotonously toward tensile direction ($\Delta g > 0$) until the cathodic limit potential of 0.26 V (SHE), while Δg in the presence of Bi^{3+} takes a maximum at about 0.55 V(SHE) and then varies toward compressive direction ($\Delta g < 0$). The value of Δg in the presence of Bi^{3+} decreases rapidly with decreasing potential in the narrow potential region between C_2 and C_3 peaks where the (2 × 2) structure is stable. Subsequently, Δg decreases linearly with decreasing potential in the potential region between 0.42 V and 0.26 V (SHE) where the $\left(p \times \sqrt{3}\right)$ – 2Bi structure is stable. There is no hysteresis of Δg between cathodic and anodic potential scans in the absence of Bi^{3+}, while some hysteresis in the presence of Bi^{3+} is observed in the potential region between 0.26 V and 0.60 V (SHE) where the Bi-UPD proceeds. The difference between Δg in the presence and absence of 5 × 10^{-3} M Bi^{3+} at –0.26 V (SHE) is $\Delta(\Delta g) = -1.35\,\mathrm{J\,m^{-2}}$which is more negative by –0.28 J m^{-2} than that ($\Delta(\Delta g) = -1.07\,\mathrm{J\,m^{-2}}$) for the Pb-UPD on the Au (111) electrode, indicating that the $\left(p \times \sqrt{3}\right)$ – 2Bi monolayer on the Au (111) electrode is subjected to large compressive stress as compared to the hcp Pb-UPD monolayer on the Au (111) electrode.

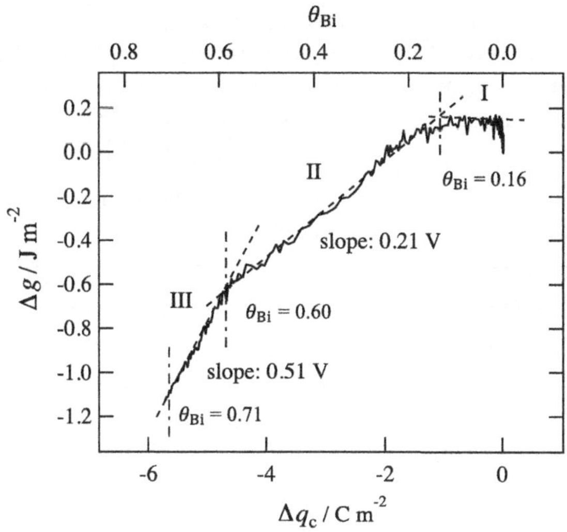

Fig. 4.8 Relationship between Δg and cathodic charge density Δq_c, which was obtained from the cyclic voltammogram and Δg vs. E curves of Fig. 4.7 measured in the cathodic potential scan in the presence of 5×10^{-3} M Bi^{3+} [14]. The surface coverage θ_{Bi} of Bi with respect to the underlying Au surface on the top of the abscissa of Fig. 4.8 is converted from the value of Δq_c by using the surface roughness $S_r = 1.2$, and the cathodic charge density $q_c = -6.66$ C m^{-2} required for Bi monolayer formation on smooth Au (111) surface. Reproduced from [14] with permission from The Electrochemical Society

Figure 4.8 shows the relationship between Δg and cathodic charge density Δq_c, which was obtained from the cyclic voltammogram and Δg vs. E curve of Fig. 4.7, measured during the cathodic potential scan in the presence of 5×10^{-3} M Bi^{3+}. The value of Δq_c is referred to zero at the anodic potential limit of 0.80 V (SHE). The net value of Δq_c was calculated by subtracting the background cathodic charge density in the absence of Bi^{3+}. The relationship between Δg and Δq_c can be classified into three (I), (II), and (III) domains: the plateau (I) domain between $\Delta q_c = 0$ and -1.3 C m^{-2} corresponding to the disorder structure, the linear (II) domain with a slope of 0.21 V between $\Delta q_c = -1.3$ and -4.8 C m^{-2} corresponding to the (2×2) structure, and the linear (III) domain with a slope of 0.51 V between -4.8 and -5.7 C m^{-2} corresponding to the $\left(p \times \sqrt{3}\right) - 2\text{Bi}$ structure. The positive slopes of 0.21 V in the linear (II) domain and of 0.51 V in the linear (III) domain are characteristic of the Bi-UPD as well as the Pb-UPD on the Au (111) electrode. The change of the positive slope from 0.21 to 0.51 V reflects sensitively on the structural change of the Bi-UPD layer from the (2×2) structure to the $\left(p \times \sqrt{3}\right) - 2\text{Bi}$ structure.

The surface coverage θ_{Bi} of Bi with respect to the underlying Au surface on the top abscissa of Fig. 4.8 is converted from the value of Δq_c by using the surface roughness $S_r = 1.2$, and the cathodic charge density $q_c = -6.66$ C m^{-2} required for the formation of $\left(p \times \sqrt{3}\right) - 2\text{Bi}$ monolayer on a smooth Au (111) surface. The (I),

(II), and (III) domains are situated in the ranges of $\theta_{Bi} = 0$ to 0.16, 0.16 to 0.60, and 0.60 to 0.71, respectively. The relationship between Δg and θ_{Bi} in Fig. 4.8 is in good agreement with that obtained from gravimetric and stress responses for the Bi-UPD on the Au (111) electrode in 1 M $HClO_4$ solution containing 2×10^{-2} M Bi^{3+} [16]. The $\left(p \times \sqrt{3}\right) - 2Bi$ structure is subjected to uniaxially electro-compressive stress along the incommensurate direction [31], while the incommensurate hcp Pb structure is subjected to biaxially electro-compressive stress [23].

Chen et al. [31] reported that in the linear (III) domain, the Bi atomic distance in the incommensurate direction of the $\left(p \times \sqrt{3}\right) - 2Bi$ structure decreases from 0.477 to 0.471 nm due to electro-compression, which reduces the Bi nearest-neighbor distance from 0.343 to 0.335 nm. If the $\left(p \times \sqrt{3}\right) - 2Bi$ monolayer is regarded as a free-standing elastic film, the relationship between $\Delta(\Delta g)$ and change in surface elastic strain $\Delta\varepsilon$ in the domain (III) is given by

$$\Delta(\Delta g) = Y_{Bi}\Delta\varepsilon d, \tag{4.21}$$

where Y_{Bi} and d are the biaxial modulus and thickness of the Bi-UPD layer, respectively. The value of $\Delta(\Delta g) = -0.55\,J\,m^{-2}$ is obtained from the net changes of Δg in the linear (III) domain. The value of $\Delta\varepsilon = -0.023$ is calculated by using the change in the Bi nearest-neighbor distance in the linear (III) domain. Furthermore, the atomic distance $d = 0.31$ nm of bulk Bi is chosen as the layer thickness. The value of $Y_{Bi} = 77.1$ GPa is eventually obtained by substituting $\Delta(\Delta g) = -0.55\,J\,m^{-2}$, $\Delta\varepsilon = -0.023$, and $d = 0.31$ nm into Eq. (4.21).

The estimated value of $Y_{Bi} = 77.1$ GPa is significantly larger than $Y_{Bi} = 47.6$ GPa (Young's modulus $E_{Bi} = 31.9$ GPa and Poisson's ratio $\nu_{Bi} = 0.33$) for polycrystalline bulk Bi. Stafford and Bertocci [16] conversely obtained $\Delta(\Delta g) = -0.37\,J\,m^{-2}$ by substituting $Y_{Bi} = 47.6$ GPa for polycrystalline bulk Bi, $\Delta\varepsilon = -0.023$, and $d = 0.34$ nm into Eq. (4.21), which is about half of that (-0.7 J m^{-2}) observed experimentally in the linear (III) domain by Stafford and Bertocci. Niece and Gewirth [35] made a potential step chrono-coulometric investigation of the Au (111) electrode in 0.1 M $HClO_4$ solution containing 2×10^{-3} M Bi^{3+} and obtained the electrocapillary curve ($\Delta\gamma$ vs. E) from the potential dependence of the surface charge density. The difference between the values of $\Delta\gamma$ in the solutions with and without 2×10^{-3} M Bi^{3+} at 0.26 V (SHE) is $\Delta(\Delta\gamma) = -0.55\,J\,m^{-2}$, which is less by a factor of 2.5 than $\Delta(\Delta g) = -1.35\,J\,m^{-2}$ at the same potential in Fig. 4.7b. As suggested by Schmickler and Leiva [36], the surface stress is directly related to the electronic structure in the metallic side of the electrode surface, which may lead to the large difference between $\Delta(\Delta g)$ and $\Delta(\Delta\gamma)$.

4.3.3 Cu-UPD on Au (111)

Roles of Co-Adsorption of Sulfate Ions

Figure 4.9 shows (a) the cyclic voltammograms and (b) the Δg vs. E curves measured at a potential scan rate of 5 mV s^{-1} for the (111)-textured Au thin-film (with a thickness of 220 nm) electrode in 0.1 M H_2SO_4 solutions with and without 10^{-3} M $CuSO_4$ [19]. The solid and dotted curves represent the results in the solutions with and without Cu^{2+}, respectively. The cathodic limit potential of 0.25 V (SHE) in Fig. 4.9 corresponds to the equilibrium potential of 10^{-3} M Cu^{2+}/Cu electrode. The anodic limit potential of 0.80 V (SHE) is chosen to avoid the onset of the surface oxygenation reaction on Au. Two pairs of cathodic and anodic peaks (C_1/A_1 and C_2/A_2) in the

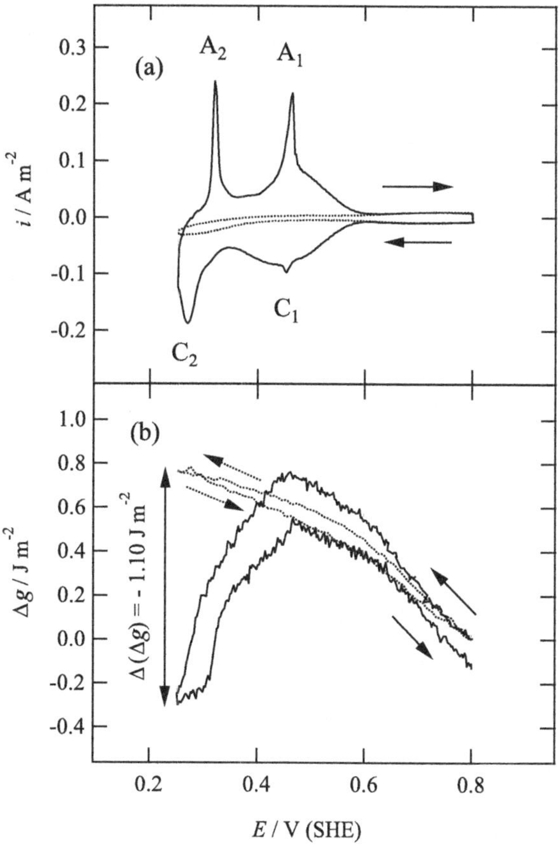

Fig. 4.9 **a** Cyclic voltammograms and **b** the Δg vs. E curves measured at a potential scan rate of 5 mV s^{-1} for the (111)-textured Au thin-film (with a thickness of 220 nm) electrode in 0.1 M H_2SO_4 solutions with and without 10^{-3} M $CuSO_4$ [19]. The solid and dotted curves represent the results in the solutions with and without Cu^{2+}, respectively. Reprinted from [19], Copyright 2007, with permission from Springer Nature

presence of Cu^{2+} are characteristic of the Cu-UPD on Au (111) electrode in sulfuric acid media, which are consistent with the results reported so far [18, 37, 38]. The Δg vs. E curve in the presence of Cu^{2+}, although some hysteresis is observed, has maxima at 0.44 V (SHE) in the cathodic potential scan and at 0.47 V (SHE) in the anodic potential scan, which correspond to the C_1 and A_1 current peaks in the cyclic voltammogram, respectively. Moreover, the Δg vs. E curve in the presence of Cu^{2+} has bending points at 0.28 V (SHE) in the cathodic potential scan and at 0.33 V (SHE) in the anodic potential scan, which correspond to the C_2 and A_2 current peaks in the cyclic voltammogram, respectively. In contrast, the Δg vs. E curve in the absence of Cu^{2+} has no maxima of Δg and no bending points; that is, Δg increases and decreases monotonously with cathodic and anodic potential scans, respectively.

The comparison between the cyclic voltammograms in the presence and absence of Cu^{2+} indicates that the Cu-UPD on Au (111) electrode starts at 0.60 V (SHE) in the cathodic potential scan. The surface stress still varies toward tensile direction, i.e., $\Delta(\Delta g) > 0$ during the cathodic potential scan from 0.60 to 0.44 V (SHE) where the Cu-UPD proceeds. In the potential region between 0.60 and 0.44 V (SHE), the STM images lacked in atomic resolution [37, 39, 40], indicating that the Cu adatoms are mobile and randomly deposited on the electrode surface. After the C_1 peak in the cathodic potential scan, the surface stress varies toward compressive direction, i.e., $\Delta(\Delta g) < 0$. The STM images [37, 38, 41] showed that the Cu-UPD layer formed at 0.40 V (SHE) in the potential region between C_1 and C_2 peaks has a honeycomb $\left(\sqrt{3} \times \sqrt{3}\right)R30°$ structure with sulfate or bisulfate ions occupying the centers of the honeycomb (see Fig. 4.10a), which was confirmed by SXS [42] and by surface-extended X-ray fine structure (SEXAFS) [43] and was also supported by electrochemical quartz crystal microbalance (EQCM) [44, 45]. The slope of the Δg vs. E curve in the presence of Cu^{2+} varies sharply toward compressive direction after the C_2 peak in the cathodic potential scan. The STM and AFM images [24, 46]

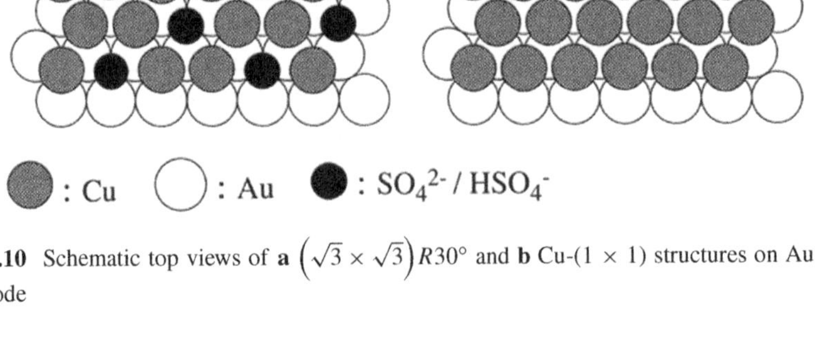

Fig. 4.10 Schematic top views of **a** $\left(\sqrt{3} \times \sqrt{3}\right)R30°$ and **b** Cu-(1 × 1) structures on Au (111) electrode

showed that the Cu-UPD layer after the C_2 peak has a pseudomorphic Cu-(1 × 1) structure (see Fig. 4.10b). Moreover, the SEXAFS [43, 47] revealed that sulfate or bisulfate ions are adsorbed on top of the Cu-(1 × 1) monolayer.

Figure 4.10 represents schematically the top views of (a) $\left(\sqrt{3}\times\sqrt{3}\right)R30°$ and (b) Cu-(1 × 1) structures on Au (111) electrode. The cathodic charge density $\Delta q_c = -4.9\,\mathrm{C\,m^{-2}}$ at 0.25 V (SHE), referred to $\Delta q_c = 0$ at 0.60 V (SHE), is calculated from the cyclic voltammogram (Fig. 4.9a) in the cathodic potential scan [19]. Assuming the surface roughness $S_r = 1.1$ of the Au (111) electrode, the value of $\Delta q_c = -4.9\,\mathrm{C\,m^{-2}}$ is close to $-4.4\,\mathrm{C\,m^{-2}}$ required for a fully discharged Cu monolayer on a smooth Au (111) electrode surface [48]. The difference between Δg in the presence and absence of 10^{-3} M Cu^{2+} at the cathodic potential limit of 0.25 V (SHE) corresponding to the Cu (1 × 1) monolayer on Au (111) electrode is $\Delta(\Delta g) = -1.10\,\mathrm{J\,m^{-2}}$, which is comparable to that ($\Delta(\Delta g) = -1.07\,\mathrm{J\,m^{-2}}$) at the cathodic limit potential of −0.24 V (SHE) corresponding to the incommensurate hcp Pb monolayer for the Pb-UPD on Au (111). Trimble et al. [17] reported that the difference between the changes in surface stress of the Au (111) in acidic sodium sulfate solutions (pH 2) with and without 10^{-3} M Cu^{2+} is $\Delta(\Delta g) = -0.6\,\mathrm{J\,m^{-2}}$ at the potential corresponding to the Cu-(1 × 1) monolayer, which is about 0.5 times as much as that in Fig. 4.9b. On the other hand, Kongstein et al. [18] reported the difference in changes in surface stress of the Au (111) electrode between 0.1 M H_2SO_4 solutions with and without 10^{-2} M Cu^{2+} is $\Delta(\Delta g) = -1.0\,\mathrm{J\,m^{-2}}$ at the potential corresponding to the Cu-(1 × 1) monolayer, which is close to that in Fig. 4.9b.

Shi and Lipkowski [49] measured the cyclic voltammograms of the Au (111) electrode in 0.1 M $HClO_4$ solutions containing 1×10^{-5} M ~ 5×10^{-3} M $Cu(ClO_4)_2$ and 10^{-3} M K_2SO_4 to calculate the electrocapillary curves from which the surface excesses Γ of Cu adatom and SO_4^{2-} ion co-adsorbed on the Au (111) electrode were obtained as a function of potential as shown in Fig. 4.11. It is remarked that $\Gamma = \frac{2}{3}$ monolayer (ML) of Cu adatom and $\Gamma = \frac{1}{3}$ monolayer (ML) of SO_4^{2-} ion at 0.36 V (SHE) in Fig. 4.11 correspond just to the honeycomb $\left(\sqrt{3}\times\sqrt{3}\right)R30°$ monolayer with SO_4^{2-} ions co-adsorbed at the centers of the honeycomb (see Fig. 4.10a). The following complicated features of the relationship between Δg and Γ in the cathodic potential scan can be drawn from the comparison between Figs. 4.9 and 4.11: (1) Δg increases toward tensile direction with decreasing Γ of SO_4^{2-} ions in the potential region between 0.80 and 0.60 V (SHE), (2) Δg increases up to the C_1 peak at 0.44 V (SHE) with increasing Γ of Cu adatom and with decreasing Γ of SO_4^{2-} ion in the potential region between 0.60 and 0.44 V (SHE), (3) Δg decreases toward compressive direction with increasing both Γ of Cu adatom and of SO_4^{2-} ion in the potential region between 0.44 and 0.36 V (SHE), and (4) Δg decreases further with increasing Γ of Cu adatom and with decreasing Γ of SO_4^{2-} ion in the potential region between 0.36 and 0.25 V (SHE) corresponding to the transition from the $\left(\sqrt{3}\times\sqrt{3}\right)R30°$ to the Cu-(1 × 1) structure. The above features demonstrate that the co-adsorption of SO_4^{2-} ions plays a vital role in the Cu-UPD on the Au (111) electrode, which reflects strongly on the magnitude and sign of Δg.

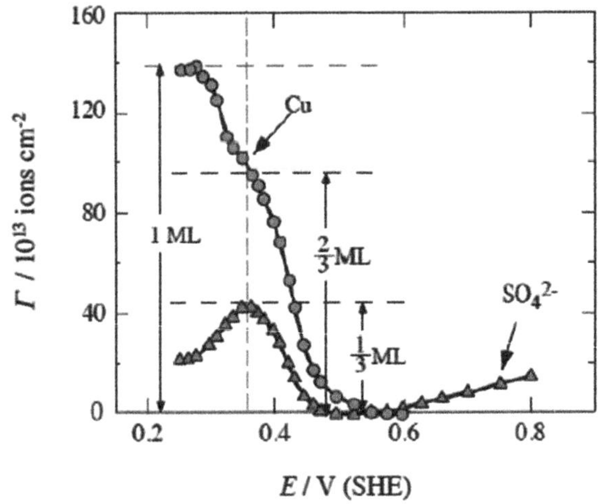

Fig. 4.11 Surface excesses Γ of Cu adatom and SO_4^{2-} ion co-adsorbed on the Au (111) surface as a function of potential, obtained from the cyclic voltammograms of the Au (111) electrode in 0.1 M $HClO_4$ solutions containing 1×10^{-5} M ~5×10^{-3} M $Cu(ClO_4)_2$ and 10^{-3} M K_2SO_4 [49]. Reprinted from [49], Copyright 1994, with permission from Elsevier

Two main factors influencing the surface stress of a solid metal electrode are the bond charge density of the electrode surface atoms and the atomic configuration of the surface layer. It is known that a clean metal surface generates usually a tensile surface stress. Ibach [50] suggested that an increased electronic charge density due to the missing bonds at a clean metal surface is redistributed to reduce the bond length between the remaining surface atoms, thereby generating the tensile surface stress at the surface. Moreover, if an adsorption species is an electron donor, the electronic charge density between the surface bonds should increase, thereby increasing the surface stress toward tensile direction. In contrast, an adsorption of electronegative species such as SO_4^{2-} and Cl^- on the surface removes the electronic charge between the surface bonds, thereby decreasing the surface stress toward compressive direction. In the cathodic potential scan from 0.80 V (SHE), the desorption of SO_4^{2-} ions proceeds predominantly prior to the onset of the Cu-UPD at about 0.60 V (SHE), and thereby Δg increases toward tensile direction. The adsorption of Cu atoms should decrease Δg toward compressive direction (i.e., $\Delta g < 0$) due to the formation of the Cu-Au surface bond. Nevertheless, Δg still increases toward tensile direction in the potential region between 0.60 and 0.44 V (SHE) despite the increase in Γ of Cu adatom, suggesting that the increase in Δg due to the desorption of SO_4^{2-} ions overcomes the decrease in Δg due to the adsorption of Cu atoms.

The mobile Cu adatoms with disorder structure which is presumed from lacking in atomic resolution of the STM image [37, 39, 40] in the potential region between 0.60 and 0.44 V (SHE) may behave like a two-dimensional gas and could not contribute to the decrease in surface stress toward compressive direction. The decrease in Δg toward compressive direction in the potential region between 0.44 and 0.36 V

(SHE) results from the co-adsorption of Cu atoms and SO_4^{2-} ions. The decrease in bond charge density between surface atoms due to the co-adsorption would bring the decrease in Δg. Moreover, the atomic configuration of co-adsorbed Cu atoms and SO_4^{2-} ions would contribute to the decrease in Δg since the honeycomb $\left(\sqrt{3} \times \sqrt{3}\right)R30°$ structure is thermodynamically stabilized by the occupation of SO_4^{2-} ions at the centers of the honeycomb. The further decrease in Δg in the potential region between 0.36 and 0.25 V (SHE) is associated with the formation of pseudomorphic Cu-(1 × 1) monolayer on the Au (111) electrode. As the lattice constant ($a_{\mathrm{Cu}} = 0.3615\,\mathrm{nm}$) of bulk Cu is less than that ($a_{\mathrm{Au}} = 0.4079\,\mathrm{nm}$) of bulk Au, the pseudomorphic Cu-(1 × 1) monolayer on Au (111) has a positive lattice mismatch of +12.8%, which would provide the increase in Δg, opposed to the experimental result.

The molecular dynamic simulations [17] indicated the increase in Δg due to the biaxial stretching of a Cu (111) slab caused by a surface strain of +12.8% is $\Delta(\Delta g) = 2.78\,\mathrm{J\,m^{-2}}$. Leiva et al. [51], using the embedded-atom method, calculated that the net change in surface stress due to the adsorption of pseudomorphic Cu-(1 × 1) monolayer on Au (111) is $\Delta(\Delta g)_{\mathrm{Cu(1\times1)/Au(111)}} = -0.28\,\mathrm{J\,m^{-2}}$, the content of which can be divided into the following two different contributions:

$$\Delta(\Delta g)_{\mathrm{Cu(1\times1)/Au(111)}} = \frac{\mathrm{d}E_{\mathrm{ads}}}{\mathrm{d}A} + \frac{\mathrm{d}U_{\mathrm{mono}}}{\mathrm{d}A}, \tag{4.22}$$

where the first contribution $\frac{\mathrm{d}E_{\mathrm{ads}}}{\mathrm{d}A} = -6.38\,\mathrm{J\,m^{-2}}$ corresponds to the change of adsorption energy of the Cu-(1 × 1) monolayer with an elastic change (dA) brought by the substrate Au (111) surface, and the second contribution $\frac{\mathrm{d}U_{\mathrm{mono}}}{\mathrm{d}A} = 6.10\,\mathrm{J\,m^{-2}}$ corresponds to the energy change due to the expansion of the isolated Cu-(1 × 1) monolayer in vacuum caused by a positive lattice mismatch. The first term in Eq. (4.22) should be negative since the formation of metallic bonds between Cu and Au becomes more favorable when the surface is expanded [51].

However, $\Delta(\Delta g)_{\mathrm{Cu(1\times1)/Au(111)}} = -0.28\ \mathrm{J\,m^{-2}}$ is significantly small as compared to $\Delta(\Delta g) = -1.10\ \mathrm{J\,m^{-2}}$ obtained experimentally at 0.25 V (SHE) corresponding to the formation of the Cu-(1 × 1) monolayer on the Au (111) electrode as shown in Fig. 4.9b. The calculation by Leiva et al. [51] has been made for the uncharged surface corresponding to E_{pzc}; thus, any changes of surface stress caused by the change of surface charge density are disregarded. Besides, the change of surface stress due to the adsorption of SO_4^{2-} ions on Cu-(1 × 1) monolayer/Au (111) is not taken into consideration for the calculation of $\Delta(\Delta g)_{\mathrm{Cu(1\times1)/Au(111)}}$. The large difference between the values of $\Delta(\Delta g)_{\mathrm{Cu(1\times1)/Au(111)}}$ obtained theoretically and experimentally may result from the neglect of the above two factors.

It is known that a $\sqrt{3} \times \sqrt{7}$ monolayer is formed by the adsorption of SO_4^{2-} ions on Cu (111) electrode at 0.175 V(SHE) [52] as well on Au (111) electrode at 1.15 V (SHE) [53]. The potential difference of about 1.0 V between the formation of the $\sqrt{3} \times \sqrt{7}$ monolayer on Cu (111) and Au (111) electrodes may result from the difference in E_{pzc} since $E_{\mathrm{pzc}} = -0.20\,\mathrm{V}$ (SHE) of the Cu (111) electrode in perchloric acid [54] is more negative by about 0.7 V than $E_{pzc} = 0.48$ V (SHE) of the unreconstructed Au (111) electrode in perchloric acid [55]. Wilms et al. [52] implied from the Moiré pattern that the formation of the $\sqrt{3} \times \sqrt{7}$ monolayer on the Cu (111) electrode expands the top Cu layer by 4%. Moreover, Trimble et al. [17] reported that the surface stress of the Cu (111) electrode varies by 0.3 J m^{-2} toward compressive direction, i.e., $\Delta(\Delta g)_{SO_4^{2-}} = -0.3$ J m^{-2}, due to the adsorption of $SO_4{}^{2-}$ ions at 0.165 V (SHE) in pH 2, 0.1 M Na_2SO_4 solution. Accordingly, the negative value of $\Delta(\Delta g)_{SO_4^{2-}}$ should be also brought by the adsorption of $SO_4{}^{2-}$ ions on the Cu-(1 × 1) monolayer/Au (111) electrode.

The adsorption of $SO_4{}^{2-}$ ions on the Cu-(1 × 1) monolayer/Au (111) electrode would be more favorable than that on the top Cu layer of the Cu (111) electrode since the Cu-(1 × 1) monolayer is more expanded than the top Cu layer of the Cu (111) electrode. The value of $\Delta(\Delta g)_{SO_4^{2-}}$ due to the adsorption of $SO_4{}^{2-}$ ions on the Cu-(1 × 1) monolayer/Au (111) electrode may be more negative than that due to the adsorption of $SO_4{}^{2-}$ ions on the Cu (111) electrode. The calculation of $\Delta(\Delta g)_{Cu(1\times1)/Au(111)}$ in Eq. (4.22) was made for the bare and unreconstructed Au (111) surface without surface charge. The value of $\Delta(\Delta g) = -1.10$ J m^{-2} at 0.25 V (SHE) in Fig. 4.9b is referred to $\Delta g = 0$ for the Au (111) electrode in the absence of Cu^{2+}. Therefore, in addition to the term of $\Delta(\Delta g)_{SO_4^{2-}}$, the correction for the changes of surface stress caused by the potential shift (i.e., the change in surface charge) from E_{pzc} to 0.25 V (SHE) for the Au (111) electrode in the absence of Cu^{2+} is needed for Eq. (4.22).

The potential of zero charge E_{pzc} of the unreconstructed Au (111) electrode in perchloric acid is 0.48 V (SHE) [55]. The addition of 10^{-3} M $SO_4{}^{2-}$ in perchloric acid shifts E_{pzc} of the unreconstructed Au (111) electrode toward negative direction at least by about 50 mV [56], which leads to $E_{pzc} = 0.43$ V (SHE) of the unreconstructed Au (111) surface in sulfuric acid. The level of surface stress for the unreconstructed Au (111) electrode at 0.43 V (SHE) corresponds to that for the reference Au (111) electrode without surface charge, which is lower by 0.13 J m^{-2} than that at 0.25 V (SHE) as seen from the dotted line in Fig. 4.9b. As a result, the correction term $\Delta(\Delta g)_{Au(111)}$ for the potential shift is −0.13 J m^{-2}. Provided that each term has an additive property, $\Delta(\Delta g)$ at 0.25 V (SHE) may be

$$\Delta(\Delta g) = \Delta(\Delta g)_{Cu(1\times1)/Au(111)} + \Delta(\Delta g)_{SO_4^{2-}} + \Delta(\Delta g)_{Au(111)}. \tag{4.23}$$

The substitution of $\Delta(\Delta g) = -1.10$ J m^{-2}, $\Delta(\Delta g)_{Cu(1\times1)/Au(111)} = -0.28$ J m^{-2}, and $\Delta(\Delta g)_{Au(111)} = -0.13$ J m^{-2} into Eq. (4.23) provides $\Delta(\Delta g)_{SO_4^{2-}} = -0.69$ J m^{-2} for the Cu-(1 × 1) monolayer/Au (111) electrode, which is 2.3 times as much as $\Delta(\Delta g)_{SO_4^{2-}} = -0.30$ J m^{-2} for the Cu (111) electrode, suggesting that the adsorption of $SO_4{}^{2-}$ on the Cu-(1 × 1) monolayer/Au (111) electrode is more energetically favorable that that on the top Cu layer of the Cu (111) electrode. Although the Cu species in the Cu-(1 × 1) monolayer on Au (111) electrode in sulfuric acid is regarded as completely discharged, X-ray absorption near-edge structure (XANES) study [47] indicated that the oxidation state of the Cu species is approximately + 1, suggesting the formation of a surface dipole in which the Cu atom has a partial positive charge, while the Au atom has a partial negative charge. The formation of the surface dipoles may promote the adsorption of SO_4^{2-} on the Cu-(1 × 1) monolayer/Au (111) electrode to decrease Δg toward compressive direction.

Roles of Co-Adsorption of chloride Ions

Figure 4.12 shows (a) the cyclic voltammogram and (b) the Δg vs. E curve measured at a potential scan rate of 5 mV s^{-1} for the (111)-textured Au thin-film (with a thickness of 220 nm) electrode in 0.1 M $HClO_4$ solution containing 10^{-3} M $Cu(ClO_4)_2$ and 10^{-3} M $KClO_4$, free from SO_4^{2-} ions [19]. The cathodic charge density $\Delta q_c = -2.4$ C m^{-2} at 0.25 V (SHE) referred to $\Delta q_c = 0$ at 0.60 V (SHE), calculated from

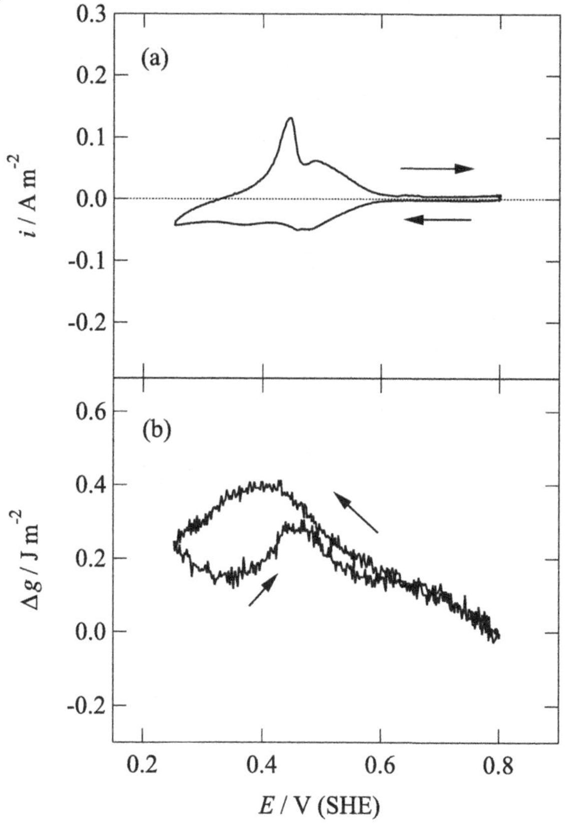

Fig. 4.12 **a** Cyclic voltammogram and **b** the Δg vs. E curve measured at a potential scan rate of 5 mV s^{-1} for the (111)-textured Au thin-film (with a thickness of 220 nm) electrode in 0.1 M $HClO_4$ solution containing 10^{-3} M $Cu(ClO_4)_2$ and 10^{-3} M $KClO_4$ free from SO_4^{2-} ions [19]. Reprinted from [19], Copyright 2007, with permission from Springer Nature

Fig. 4.12a, is only 50% of $\Delta q_c = -4.9$ C m^{-2} (calculated from Fig. 4.9a) required for the formation of the Cu-(1 × 1) monolayer on Au (111) electrode, indicating that the Cu-UPD on Au (111) electrode is strongly hindered in perchloric acid media without SO_4^{2-} ions. In response to the cyclic voltammogram, the change in surface stress in the potential range from 0.40 to 0.25 V (SHE) in the cathodic potential scan in Fig. 4.12b is $\Delta(\Delta g) = -0.17$ J m^{-2}, which is significantly small as compared to $\Delta(\Delta g) = -0.90$ J m^{-2} in the same potential range in Fig. 4.9b, indicating that the formation of Cu-(1 × 1) monolayer is not accomplished. The feature of the cyclic voltammogram in Fig. 4.12a is consistent with that of the cyclic voltammogram obtained for the Cu-UPD on Au (111) in 0.1 M $HClO_4$ solution containing 10^{-2} M $Cu(ClO_4)_2$ [57].

Hotlos et al. [57] pointed out that the broad cathodic and anodic peaks in the cyclic voltammogram are associated with trace amounts of Cl^- ions ($\leq 10^{-6}$ M) contained as impurity in the solution. The STM images [57] of the Au (111) electrode surface in 0.1 M $HClO_4$ solution containing 10^{-2} M $Cu(ClO_4)_2$ with small amounts of Cl^- ions (10^{-6} M) showed that a "(5 × 5)" structure is transformed into a (2 × 2) superstructure upon changing the potential from 0.55 to 0.38 V (SHE). Moreover, the careful measurements revealed that the "(5 × 5)" structure is not a real (5 × 5) but an incommensurate structure between (4 × 4) and (5 × 5). On the other hand, in the case where the concentration of Cl^- ions exceeds a critical level ($>10^{-5}$ M), the "(5 × 5)" phase remains stable in the Cu-UPD potential range up to the onset of the Cu bulk deposition, while the (2 × 2) phase is unstable and dissolves. Hotlos et al. [57] proposed that the "(5 × 5)" structure consists of a bilayer of Cu and Cl, in which each Cl^- ion is adsorbed in a hollow site between three Cu adatoms with Cu-Cu and Cl–Cl interatomic distances of 0.367 nm. The results of SEXAS by Wu et al. [58] supported the above Cu-Cl bilayer model. The interatomic distance of 0.367 nm in the Cu-Cl bilayer model is shorter by 4% than that (0.382 nm) in the (111) plane of CuCl crystal with a zinc blende lattice.

Figure 4.13 shows (a) the cyclic voltammogram and (b) the Δg versus E curve measured at a potential scan rate of 5 mV s^{-1} for the (111)-textured Au thin-film (with a thickness of 220 nm) electrode in 0.1 M $HClO_4$ solution containing 10^{-3} M $Cu(ClO_4)_2$ and 10^{-3} M KCl [19]. It is distinct from the comparison between Figs. 4.12a and 4.13a that the Cu-UPD process on the Au (111) electrode is significantly enhanced by an addition of 10^{-3} M Cl^- ions in perchloric acid media. Two pairs of cathodic and anodic peaks (C_1/A_1 and C_2/A_2) in Fig. 4.13a are characteristic of the Cu-UPD on the Au (111) electrode in perchloric acid or acidic perchlorate media containing 10^{-4} M or 10^{-3} M Cl^- ions, which are consistent with the previous results [57, 58]. The potential at the C_1 peak in Fig. 4.13a is 0.52 V (SHE) which is more positive by about 80 mV than that (0.44 V (SHE)) at the corresponding C_1 peak in Fig. 4.9a, suggesting that the co-adsorption of Cu with Cl^- ions is stronger than that with SO_4^{2-} ions as reported by Shi et al. [59]. In response to the enhancement of the Cu-UPD on the Au (111) electrode due to the addition of Cl^- ions, the surface stress varies significantly depending on the potential region as shown in Fig. 4.13b.

The value of Δg increases toward tensile direction with decreasing potential from 0.80 (SHE) to 0.65 V (SHE), which is associated with the desorption of Cl^- ions

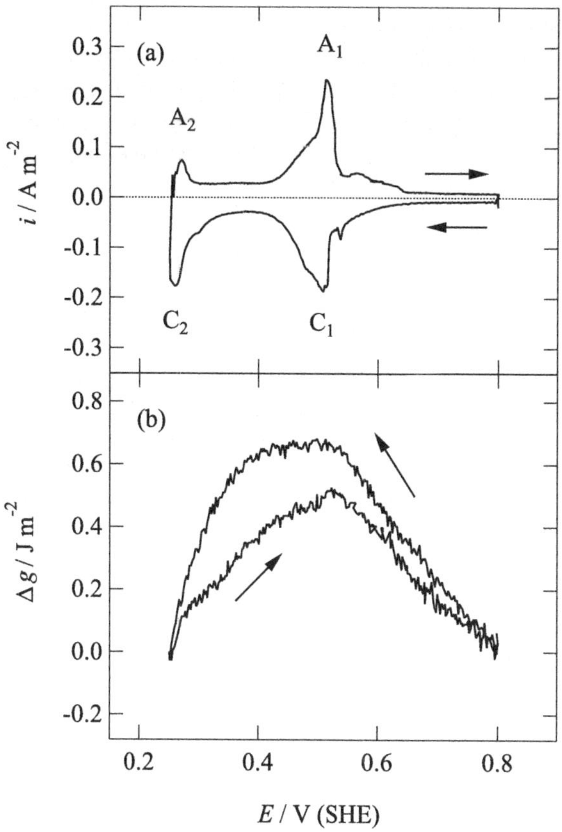

Fig. 4.13 **a** Cyclic voltammogram and **b** the Δg vs. E curve measured at a potential scan rate of 5 mV s^{-1} for the (111)-textured Au thin-film (with a thickness of 220 nm) electrode in 0.1 M $HClO_4$ solution containing 10^{-3} M $Cu(ClO_4)_2$ and 10^{-3} M KCl [19]. Reprinted from [19], Copyright 2007, with permission from Springer Nature

as supported by the chrono-coulometric results [58]. Both surface excesses of Cu adatom and Cl^- ion increase with decreasing potential from 0.65 to 0.52 V due to the co-adsorption [58], while Δg increases continuously down to 0.52 V (SHE) corresponding to the C_1 peak. The STM images [57] indicated that the ordered "(5 × 5)" structure is formed at potentials more negative than 0.52 V (SHE). No clear STM images of the ordered CuCl bilayer structure were observed at potentials more positive than 0.52 V (SHE), suggesting that the adlayer formed in the potential region between 0.62 and 0.52 V (SHE) has a disordered structure.

Although the reason for the increase in Δg is not made clear, an attractive lateral interaction may work between Cu and Cl species in the disordered adlayer to provide the increase in Δg. The value of Δg keeps nearly constant in the cathodic potential scan from 0.52 to 0.40 V (SHE). At potentials more negative than 0.52 V (SHE), the disordered adlayer is transformed to the ordered CuCl bilayer. The formation of the ordered CuCl bilayer would lead to the decrease in Δg toward compressive direction which is caused by a lateral expansion due to the strong bond between Cu and Cl atoms. The stagnant value of Δg in the cathodic potential scan from 0.52 to 0.40 V (SHE) may be brought by balancing the formation of the disordered adlayer with the transformation to the ordered CuCl bilayer. The continuous decrease in Δg in the cathodic potential scan from 0.40 to 0.25 V (SHE) results from the growth of the CuCl bilayer.

4.3.4 Pd-UPD on Au (111)

Figure 4.14 shows (a) the cyclic voltammograms and (b) the Δg vs. E curves measured by changing the cathodic limit potential and fixing the anodic limit potential (0.53 V (SSE)) at a potential scan rate of 5 mV s^{-1} for the (111)-textured Au thin-film (with a thickness of 250 nm) electrode in 0.1 M H_2SO_4 solution containing 10^{-3} M H_2PdCl_4 [20]. In Fig. 4.14a, the increase in cathodic current density from about 0.28 V (SSE) in the cathodic potential scan corresponds to the progress in Pd-UPD on Au (111) electrode. In the case where the cathodic limit potential is more positive than 0.08 V (SSE), the anodic current peak appears at about 0.23 V (SSE) in the anodic potential scan, which results from the anodic stripping of the Pd-UPD layer. When the cathodic limit potential is more negative than 0.08 V, the additional anodic current peak appears in the potential range of 0.12 – 0.16 V (SSE), which results from the anodic stripping of the bulk-deposited Pd layer.

A rapid increase in cathodic current density from 0.08 V (SSE) in the cathodic potential scan corresponds to the progress in Pd-bulk deposition (overpotential deposition of Pd: Pd-OPD). The boundary potential between Pd-UPD and Pd-OPD is not clearly distinct, but it is located at about 0.08 V (SSE). The potential region between 0.08 and 0.28 V (SSE) where the Pd-UPD proceeds on the Au (111) electrode is consistent with that reported by Baldauf and Kolb [60]. An adsorbed $[PdCl_4]^{2-}$ layer would be present on the surface in the Pd-UPD potential region since E_{pzc} for the Au (111) electrode in sulfuric acid is about −0.2 V (SSE) [61]. The deposition of Pd is presumed to occur via reduction of the adsorbed tetrachloro palladate ($[PdCl_4]^{2-} + 2e^- \rightarrow Pd + 4Cl^-$). The small coupled anodic/cathodic peaks at about 0.49 V (SSE) in Fig. 4.14 are associated with the replacement of adsorbed $[PdCl_4]^{2-}$ by Cl^- [61, 62].

In Fig. 4.14b, the anodic limit potential is fixed at 0.38 V (SSE), while the cathodic limit potential is varied in turn toward negative direction. The surface stress increases toward tensile direction, i.e., $\Delta g > 0$, in the cathodic potential scan from 0.38 V (SSE) at which Δg is referred to zero. The significant increase in Δg is observed at potentials more negative than 0.08 V (SSE), suggesting that the tensile stress in the bulk-deposited Pb layer is larger than that in the Pd-UPD monolayer. The surface stress transients in the anodic potential scan from the cathodic limit potential exhibit a large hysteresis with negative shift of the cathodic limit potential, which is caused by a slow growth process of the Pd-bulk deposition. Nonetheless, the Δg vs. E curves

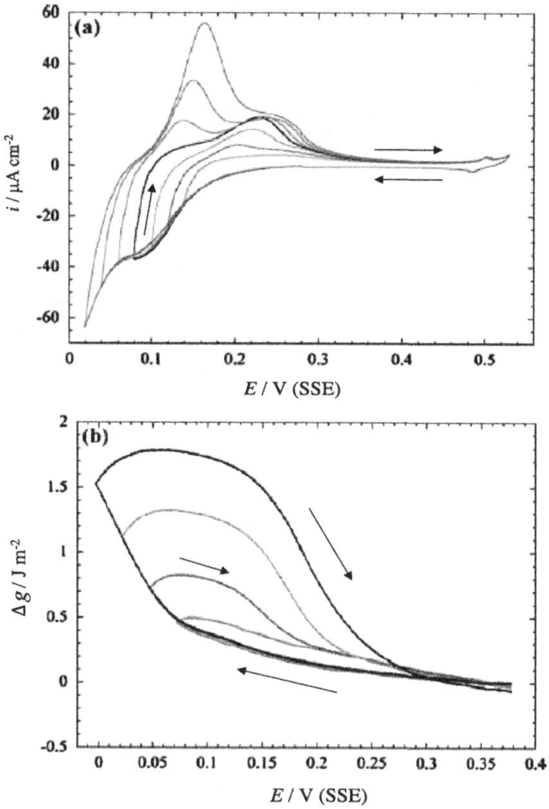

Fig. 4.14 **a** Cyclic voltammograms and **b** the Δg vs. E curves measured by changing the cathodic limit potential and fixing the anodic limit potential (0.53 V (SSE)) at a potential scan rate of 5 mV s^{-1} for the (111)-textured Au thin-film (with a thickness of 250 nm) electrode in 0.1 M H_2SO_4 solution containing 10^{-3} M H_2PdCl_4 [20]. Reprinted with the permission from [20], Copyright 2009, American Chemical Society

merge in the cathodic potential scan after the potential was returned to the anodic limit potential of 0.38 V (SSE), indicating that the surface alloying between Pd and Au atoms does not occur in the examined potential region during experiments [20].

The Δg vs. E curves in Fig. 4.14b do not have a stress maximum although a stress maximum in the Δg vs. E curves has been observed for the Cu-UPD [17–19] or Al-UPD [63] on the Au (111) electrode, known as a pseudomorphic UPD system with positive lattice mismatch (expansion of adsorbate lattice) between the adsorbate and substrate. Despite the progress in Pd-UPD on the Au (111) electrode, the surface stress increases continuously toward tensile direction, which is opposite to the decrease in surface stress toward compressive direction observed for the Cu-UPD [17–19] or Al-UPD [63] on the Au (111) electrode. Figure 4.15 shows the average changes in surface stress (in the cathodic and anodic stress response) $\overline{\Delta g}$ as a function of the net cathodic charge density Δq_c obtained for a series of potential

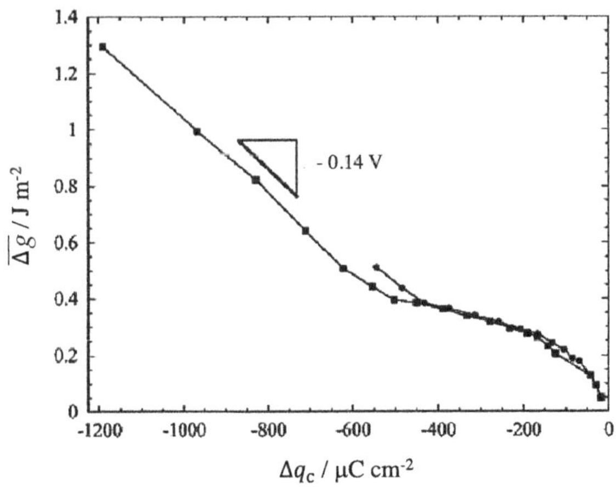

Fig. 4.15 Average surface stress change (in the cathodic and anodic stress response) $\overline{\Delta g}$ as a function of the net cathodic charge density Δq_c obtained for a series of potential step from 0.38 V (SSE) down to various potentials [20]. The data points marked with ● and ■ in Fig. 4.15 correspond to the results of two separate experiments performed in the narrow (UPD) and wide potential (OPD) regions, respectively. Reprinted with the permission from [20], Copyright 2009, American Chemical Society

step from 0.38 V (SSE) down to various potentials for the (111)-textured Au thin-film (with a thickness of 250 nm) electrode in 0.1 M H_2SO_4 solution containing 10^{-3} M H_2PdCl_4 [20]. The data points marked with ● and ■ in Fig. 4.15 correspond to the results of two separate experiments performed in the narrow (UPD) and wide potential (OPD) regions, respectively. At the potential step to the Pd-UPD region (more positive than 0.08 V), the surface stress increases toward tensile direction and subsequently keeps a steady-state value, depending on the step-down potential. When the potential is reversed back to 0.38 V (SSE), the surface stress returns to its original value, indicating that the surface stress change due to the Pd-UPD is reversible.

In contrast, at the potential step to the Pd-OPD region, the surface stress jumps initially toward tensile direction and increases continuously, depending on the step-down potential, which reflects the steady-state growth of the bulk-deposited Pd layer in the Pd-OPD region. In Fig. 4.15, the value of $\overline{\Delta g}$ is about 0.4 J m^{-2} at $\Delta q_c = -4.5$ C m^{-2} corresponding to the cathodic charge density required for the formation of the Pd-UPD monolayer and it increases linearly with a slope of −0.14 V in the Pd-OPD region ($\Delta q_c < -4.5$ C m^{-2}). Since the lattice parameters for bulk Pd and Au are $a = 0.389$ and 0.408 nm, respectively, the Pd-UPD layer has +4.9% lattice mismatch with respect to the Au substrate. The contribution of the lattice mismatch to the surface stress for the pseudomorphic (1 × 1) Pd-UPD layer on the Au (111) electrode [20] may be estimated by

$$\Delta g = \frac{E_{\mathrm{Pd}(111)}}{1 - \nu_{\mathrm{Pd}(111)}} \varepsilon_{\mathrm{mf}} d_{\mathrm{Pd}(111)} n, \tag{4.24}$$

where $E_{Pd(111)}$ is Young's modulus for the Pd (111) layer, $\nu_{Pd(111)}$ is Poisson's ratio, $\varepsilon_{mf} = 0.049$ is the misfit strain, $d_{Pd(111)} = 0.225$ nm is the thickness of the Pd-UPD monolayer, and n is the number of the monolayers on the surface. The values of $E_{Pd(111)} = 136$ GPa and $\nu_{Pd(111)} = 0.528$ are calculated from the elastic compliances of Pd [28]. The value of $\Delta g = 3.17$ J m^{-2} is estimated as the contribution of the lattice mismatch by inserting these values into Eq. (4.24) for the Pd-UPD monolayer, which is larger by about one order of magnitude than that ($\Delta g = 0.4$ J m^{-2}) obtained experimentally in the Pd-UPD region in Fig. 4.15.

The formation of metallic bonds between Pd and Au atoms contributes to the surface stress change toward compressive direction ($\Delta g < 0$) which counterbalances the surface stress change toward tensile direction due to the positive lattice mismatch. Equation (4.22) derived by Leiva et al. [51] can be applied to the (1×1) Pd/Au (111) system since the Pd-UPD layer on the Au (111) electrode has the same pseudomorphic structure as the Cu-UPD layer on the Au (111) electrode. The value of $\Delta(\Delta g)_{pd(1\times1)/Au(111)} = -0.1$ J m^{-2} calculated by Leiva et al. [51] is close to the experimental value of $\Delta g = 0.4$ J m^{-2} as compared to that ($\Delta g = 3.17$ J m^{-2}) due to the lattice mismatch calculated from Eq. (4.24). In the case of the Cu-UPD on the Au (111) electrode, the desorption of anion species and followed by the co-adsorption of anion species with Cu atoms influences the magnitude and direction of surface stress change. Particularly, the co-adsorption of $SO_4{}^{2-}$ ion on the Cu adlayer or the co-adsorption of Cl^- ion followed by the formation of the CuCl bilayer on the Au (111) electrode induces the compressive surface stress, which provides the surface stress maximum during the Cu-UPD process.

On the other hand, in the case of the Pd-UPD on the Au (111) electrode, no surface stress maximum is observed during the Pd-UPD process. Since E_{pzc} of Pd electrode is more negative than that of Au electrode [5], the adsorbed $[PdCl_4]^{2-}$ is not desorbed in the Pd-UPD potential region and remains on the Pd-UPD monolayer. The adsorbed $[PdCl_4]^{2-}$ on the Pd-UPD monolayer may contribute to the decrease in surface stress toward compressive direction. However, its contribution would not be enough to induce a surface stress maximum. The surface stress change of $\Delta g = 0.62$ J m^{-2} for the Pd-OPD monolayer (corresponding to $\Delta q_c = -9.0$ C m^{-2}) is obtained from the slope of -0.14 V in the Pd-OPD region of Fig. 4.15. The value of Δg for the Pd-OPD monolayer is larger by about 50% than that (0.4 J m^{-2}) for the Pd-UPD monolayer, suggesting that the contribution of the decrease in surface stress toward compressive direction due to the bonding between Pd and Au atoms diminishes with growth of the Pd-OPD layer [20].

The changes in surface stress during Pd deposition on the Au (111) electrode in 0.1 M H_2SO_4 solution containing 10^{-3} M $PdSO_4$ were also measured to examine the contribution of $[PdCl_4]^{2-}$ to the changes in surface stress [20]. The surface stress response during Pd deposition in the solution containing $PdSO_4$ was similar to that in the solution containing $[PdCl_4]^{2-}$. The adsorbed $SO_4{}^{2-}$ as well as $[PdCl_4]^{2-}$ is not desorbed in the Pd-UPD potential region and remains on the Pd-UPD monolayer. Nevertheless, the anodic stripping of the deposited Pd is kinetically hindered in chloride-free solution [20]. The in situ gravimetry by EQCM combined with cyclic voltammetry and surface stress measurement in the potential range covering from Pd-UPD to Pd-OPD on Au (111) electrode in the solution containing $PdSO_4$ [20] confirmed that the anodic stripping of the deposited Pd is incomplete in the absence of chloride in solution and some part of the Pd layer remains on the surface, indicating that the adsorbed $[PdCl_4]^{2-}$ species promotes the anodic stripping of the deposited Pd.

4.4 Surface Alloying

In spite of the immiscibility of foreign metal in bulk metal, the surface alloy formation limited to surface monolayer has been observed in many adsorbate/substrate systems [64] such as Pb, Sn, or Sb on Cu (*hkl*) and Pb or Sn on Ni (*hkl*) that were prepared with vapor deposition in vacuum. In addition, the surface alloy formation has been also observed in several UPD systems such as Pb or Tl on both Au (111) and Ag (111) [25, 65]. Various ion scattering studies by low-energy alkali ion (500 eV Li^+) [66–69], coaxial impact-collision ion (3 keV Ne^+) [70], and medium-energy ion (100 keV H^+) [71] revealed that the magnitude of surface rumpling for the surface alloy phases prepared with vapor deposition in vacuum is smaller than that predicted from the atomic radii in bulk metals based on a hard sphere model in the case where the alloying adsorbate atoms are larger than the substrate atoms in the surface alloy phases such as Ni (111) $\left(\sqrt{3}\times\sqrt{3}\right)R30°$-Pb and Cu (111) $\left(\sqrt{3}\times\sqrt{3}\right)R30°$-Sn. The exceptional case is the surface alloy of W (100) $c(2\times2)$-Cu for which the alloying adsorbate Cu atom is smaller than the substrate W atom.

The nearest-neighbor interatomic distance $d_{\mathrm{B-A}}$ in the surface alloy (i.e., the B–A distance within the surface alloy phase formed by deposition of B on A) can be determined from the magnitude of the surface rumpling by taking the atomic configuration of the uppermost surface alloy monolayer and the underneath substrate layer into account. The calculated $d_{\mathrm{B-A}}$ for the surface alloys in which the alloying adsorbate atoms are larger than the substrate atoms is also shorter than that predicted from the hard sphere model, indicating that the effective radius of the adsorbate B atom is reduced by surface alloying if the radius of the substrate A atom is assumed to be unchanged. For example, $d_{\mathrm{Pb-Ni}}$ in the surface alloy of Ni (111) $\left(\sqrt{3}\times\sqrt{3}\right)R30°$-Pb is 0.258 ± 0.001 nm [71, 72] which is significantly shorter than that ($d_{\mathrm{Pb-Ni}} = 0.300$ nm) estimated from the sum of atomic radii (0.1246 nm for Ni and 0.1750 nm for Pb) in the respective bulk metals based on a simple hard sphere model, and the effective radius of Pb in the surface alloy obtained under the assumption that the atomic radius of Ni is unchanged, is 0.133 nm, which is smaller by about 25% than the atomic radius of Pb in bulk Pb.

The single-bond covalent radii of Ni and Pb atoms are 0.115 nm and 0.154 nm, respectively, as defined by Pauling [73]. The value of $d_{\mathrm{Pb-Ni}} = 0.258\,\mathrm{nm}$ in the surface ally is close to $d_{\mathrm{Pb-Ni}} = 0.269\,\mathrm{nm}$ estimated from the covalent radii of Pb and Ni atoms, suggesting that the Pb-Ni bond in the surface alloy has a covalent character [64]. In addition, in the case where the radius of the adsorbate B atom is larger than that of the substrate A atom, the interatomic distance d_{B-A} in the surface alloy is shorter than that in the corresponding bulk alloy, which has been confirmed by the density functional theory (DFT) calculations [74]. For example, $d_{Sn-Ni} = 0.253$ nm in the surface alloy of Ni (111) $\left(\sqrt{3} \times \sqrt{3}\right)R30°$-Sn [67] is shorter than $d_{Sn-Ni} = 0.261 \sim 0.264$ nm in the bulk Ni_3Sn alloy [64]. Furthermore, the DFT calculations [75] for the surface structure and surface stress of a series of different surface alloy phases confirmed that the effective radius of adsorbate atom is reduced by alloying for the systems with larger adsorbate atoms, and showed that the tensile stress is reduced by alloying and the surface stress becomes compressive for some systems in which the radius of adsorbate atom is significantly larger than that of substrate atom. By contrast, the tensile stress is increased by alloying for the surface phase of W (100)$c(2 \times 2)$-Cu in which the adsorbate Cu atom is smaller than the substrate W atom.

Figure 4.16 shows (a) the changes in surface stress Δg due to alloying and (b) the surface stress g of surface alloy, plotted versus the calculated difference in atomic radius Δr between adsorbate and substrate atoms for the surface alloys in which the radius of adsorbate atom is larger than that of the substrate atom [75]. In Fig. 4.16, it is reminded that the results for the surface alloys in which the radius of adsorbate atom is smaller than that of substrate atom are excluded from the original figure in ref [75] to avoid the complexity. The value of Δg in Fig. 4.16a was obtained by subtracting the absolute value of surface stress, calculated for the respective clean substrate metal surface from the absolute value of surface stress g, calculated for the respective surface alloy in Fig. 4.16b. It is clear from Fig. 4.16a that the changes in surface stress due to alloying tend to shift toward more compressive direction ($\Delta g < 0$) with increasing Δr. It is noticed that the unit of the abscissa in Fig. 4.16a and b is expressed by Ångström (Å = 0.1 nm).

On a clean metal surface, because of the missing bond, the electronic charge accumulates between the surface atoms to cause the contraction of the equilibrium bond distance between surface atoms, by which the tensile stress arises on the clean metal surface [50]. If the electronic charge accumulated between surface atoms is removed due to the bonding of surface atoms with adsorbate atoms, the surface stress varies toward compressive direction. Besides, in the case where the radius of the adsorbate atom is significantly larger than that of the substrate atom, the substitution of the substrate atoms in the surface layer by the adsorbate atoms, followed by the surface alloy formation, induces the large changes in surface stress toward compressive direction due to the large difference in atomic radius between adsorbate and substrate atoms. However, the compressive surface stress may be released to some extent by achieving shorter interatomic distance (i.e., significant reduction in the effective radius of adsorbate atom) in the resulting surface alloy phase, which may be energetically favorable for the formation of the surface alloy in such systems.

By contrast, in the case where the radius of the adsorbate atom is smaller than that of the substrate atom such as W (100) $c(2 \times 2)$-Cu, the surface alloy formation increases the tensile surface stress. Unfortunately, there have been no reports of the

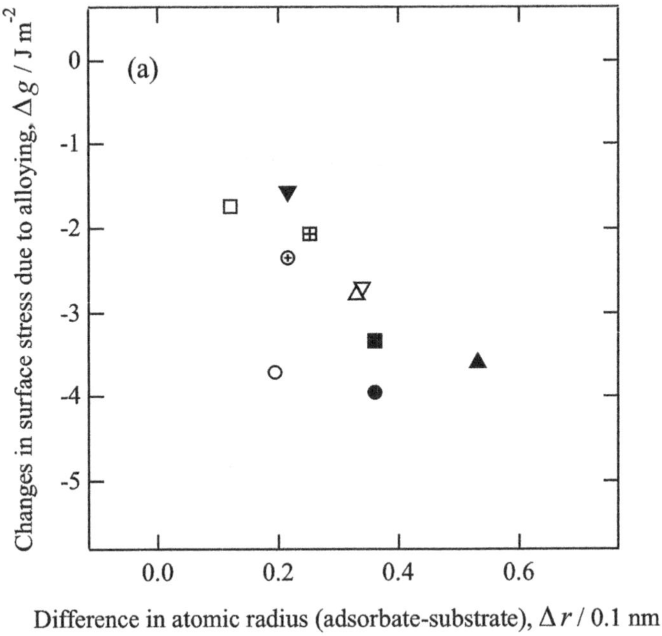

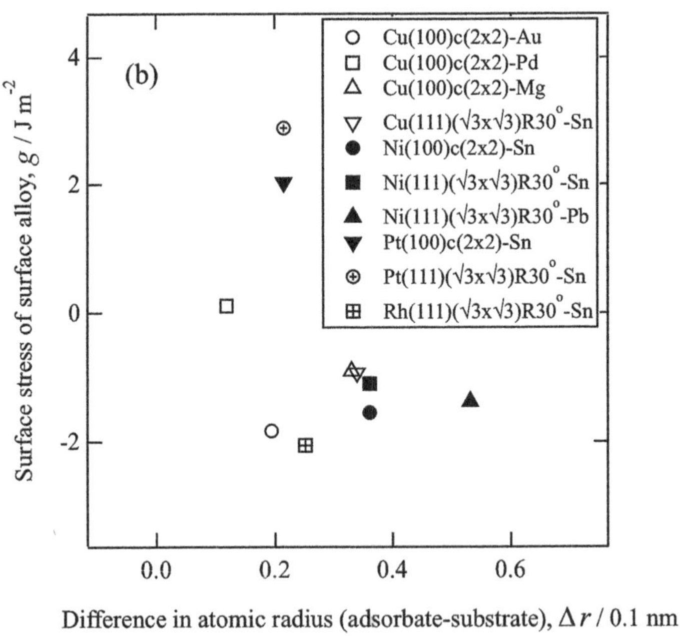

Fig. 4.16 **a** The changes in surface stress Δg due to alloying and **b** the surface stress g of surface alloy, plotted versus the calculated difference in atomic radius Δr between adsorbate and substrate atoms for the surface alloys in which the radius of adsorbate atom is larger than that of the substrate atom [75]. The unit of the abscissa in Fig. 4.16a and b is expressed by Ångström (Å = 0.1 nm). Modified from [75], Copyright 2004, with permission from Elsevier

experimental results of changes in surface stress for the formation of the surface alloys with a large difference in atomic radius between adsorbate and substrate atoms such as Cu (111) $\left(\sqrt{3}\times\sqrt{3}\right)$ $R30°$-Sn and Ni (111) $\left(\sqrt{3}\times\sqrt{3}\right)$ $R30°$-Pb, although the experimental value of $\Delta g \approx -5\,\mathrm{J\,m^{-2}}$ [50] has been obtained at the deposition of Ag monolayer (but the formation of surface alloy was not confirmed) on Pt (111) in which the radius of Ag atom is larger by 4.3% than that of Pt atom.

The corrosion inhibition of ferrous metals such as steels, nickel, and iron in the presence of Pb^{2+} or Sn^{2+} in perchloric acid or acidic perchlorate solution has been ascribed to the UPD of Sn or Pb on the substrate metal [76]. The in situ EXAFS analysis of the Pb-UPD on surface-roughened (*sr*-) polycrystalline Ni electrode in acidic perchlorate solution containing Pb^{2+} [77] suggested that the Pb-UPD layer consists of a mixture of the surface alloy and overlayer, and indicated that the average interatomic distance $d_{\mathrm{Pb-Ni}}$ in the Pb-UPD layer is 0.264 nm, which is significantly small as compared to that ($d_{\mathrm{Pb-Ni}} = 0.300$ nm) estimated from a hard sphere model. If the atomic radius (0.1246 nm) of Ni is unchanged, the effective radius of Pb atom in the Pb-UPD layer is 0.139 nm, which is smaller by 21% than that (0.175 nm) of Pb atom in bulk Pb.

Similarly, in situ EXAFS analysis of the Sn-UPD on the *sr*-polycrystalline Ni electrode in perchloric acid containing Sn^{2+} [78] suggested that Ni atoms at face-centered cubic (fcc) sites in the first Ni layer are substituted by Sn atoms like in a surface alloy, and indicated that the average interatomic distance $d_{\mathrm{Sn-Ni}}$ in the Sn-UPD layer is 0.256 nm, which is significantly small as compared to that ($d_{\mathrm{Sn-Ni}} =$ 0.265 nm or 0.276 nm) estimated from the atomic radii ($r_{\alpha-\mathrm{Pb}} = 0.145$ nm or $r_{\beta-\mathrm{Pb}} =$ 0.1551 nm) of metal atoms in the corresponding bulk metals based on a hard sphere model. The effective radius of Sn atom in the Sn-UPD layer is 0.131 nm, which is smaller by 6.7% or 13.3% than that (0.1405 nm or 0.1551 nm) of Sn atom in bulk α- or β-Sn. The significant reduction in the effective radius of Pb or Sn adsorbate atom has been observed in the Pb- or Sn-UPD layer on the *sr*-polycrystalline Ni electrode in aqueous solution as well as the surface alloys prepared with vapor deposition in vacuum.

Although the measurement of changes in surface stress during the Pb-UPD or Sn-UPD on a smooth Ni electrode has not been achieved, a large compressive surface stress should arise during the UPD process. The compressive surface stress may be relaxed by the reduction in effective radius of the adsorbate atom in the UPD layer due to surface alloying, which may contribute to the minimization of total energy for the equilibrium structure of the UPD layer.

References

1. Kolb DM (1978) Physical and electrochemical properties of metal monolayers on metallic substrates. In: Gerischer H, Tobias CW (eds) Advances in electrochemistry and electrochemical engineering. vol 11. Wiley, New York, pp 125–271

2. Adzic RR (1984) Electrocatalytic properties of the surfaces modified by foreign metal adatoms. In: Gerischer H (ed) Advances in electrochemistry and electrochemical engineering. vol 13. Wiley, New York, pp 159–260
3. Kolb DM, Przasnyski M, Gerischer H (1974) J Electroanal Chem 54:25–38
4. Trasatti S (1975) Zeit Phys Chem NF 98:75–94
5. Trasatti S (1977) The work function in electrochemistry. In: Gerischer H, Tobias CW (eds) Advances in electrochemistry and electrochemical engineering. vol 10. Wiley, New York, pp 213–321
6. Lang ND, Kohn W (1971) Phys Rev B 3:1215–1223
7. Tsukada M (1989) Introduction to surface physics (Japanese). Tokyo Univ Press, Tokyo, pp 130
8. Langmuir I (1918) J Am Chem Soc 40:1361–1403
9. Frumkin AN (1925) Z Phys Chem 116 U:466–484
10. Brunt TA, Rayment T, O'Shea SJ, Welland ME (1996) Langmuir 12:5942–5946
11. Friesen C, Dimitrov N, Cammarata RC, Sieradzki K (2001) Langmuir 17:807–815
12. Seo M, Yamazaki M (2004) J Electrochem Soc 151:E276–E281
13. Stafford GR, Bertocci U (2007) J Phys Chem C 111:17580–17586
14. Seo M, Yamazaki M (2008) ECS Trans 11:13–21
15. Shin JW, Bertocci U, Stafford GR (2010) J Phys Chem C 114:7926–7932
16. Stafford GR, Bertocci U (2006) J Phys Chem B 110:15493–15498
17. Trimble T, Tang L, Vasilljevic N, Dimitrov N, van Schilfgaarde M, Friesen C, Thompson CV, Seel SC, Floro JA, Sieradzki K (2005) Phys Rev Lett 95:166106-1-4
18. Kongstein OE, Bertocci U, Stafford GR (2005) J Electrochem Soc 152:C116-C123
19. Seo M, Yamazaki M (2007) J Solid State Electrochem 11:1365–1373
20. Stafford GR, Bertocci U (2009) J Phys Chem C 113:261–268
21. Kolb DM (1996) Prog Surf Sci 51:109–173
22. Haiss W, Nichols RJ, Sass JK, Charle KP (1998) J Electroanal Chem 452:199–202
23. Toney MF, Gordon JG, Samant MG, Borges GL, Melroy OR, Yee D, Sorensen LB (1995) J Phys Chem 99:4733–4744
24. Green MP, Hansen KJ (1991) Surf Sci Lett 259:L743–L749
25. Canal D, Oden PI, Müller U, Schmidt E, Siegenthaler H (1995) Electrochim Acta 40:1223–1235
26. Nagl C, Haller O, Platzgummer E, Schmid M, Varga P (1994) Surf Sci 321:237–248
27. Ibach H (2006) Physics of surfaces and interfaces. Springer, Berlin, pp 131
28. Brandes EA, Brook GB (eds) (1992) Smithells metals reference book. 7th edn. Butterworth-Heinemann, Oxford, pp 15–16
29. Koch R (1994) J Phys Condens Matter 6:9519–9550
30. Shin JW, Bertocci U, Stafford GR (2010) J Phys Chem C 114:17621–17628
31. Chen C-H, Kepler KD, Gewirth AA, Ocko BM, Wang J (1993) J Phys Chem 97:7290–7294
32. Chen C-H, Gewirth AA (1992) J Am Chem Soc 114:5439–5442
33. Jeffrey CA, Harrington DA, Morin S (2002) Surf Sci Lett 512:L367–L372
34. Hara M, Nagahara Y, Inukai J, Yoshimoto S, Itaya K (2006) Electrochim Acta 51:2327–2332
35. Niece BK, Gerwirth AA (1996) Langmuir 12:4909–4913
36. Schmickler W, Levia E (1998) J Electroanal Chem 453:61–67
37. Hachiya T, Honbo H, Itaya K (1991) J Electroanal Chem 315:275–291
38. Batina N, Will T, Kolb DM (1992) Faraday Discuss 94:93–106
39. Magnussen OM, Hotlos J, Nichols RJ, Kolb DM, Behm RJ (1990) Phys Rev Lett 64:2929–2936
40. Magnussen OM, Behm RJ (1999) J Electroanal Chem 467:258–269
41. Green MP, Hansen KP (1992) J Vac Sci Technol, A 10:3012–3018
42. Toney MF, Howard JN, Richer J, Borges GL, Gordon JG II, Melroy OR (1995) Phys Rev Lett 75:4472–4476
43. Melroy OR, Samant MG, Borges GL, Gordon JG II, Blum L, White JH, Albarelli MJ, McMillan M, Abruna HD (1988) Langmuir 4:728–732
44. Borges GL, Kanazawa KK, Gordon JG II, Ashley K, Richer J (1994) J Electroanal Chem 360:281–284

45. Watanabe M, Uchida H, Miura M, Ikeda N (1995) J Electroanal Chem 384:191–195
46. Ikemiya N, Miyaoka S, Hara S (1994) Surf Sci 311:L641–L648
47. Tadjeddine A, Tourillon G, Guay D (1991) Electrochim Acta 36:1859–1862
48. Herrero E, Buller LJ, Abruna HD (2001) Chem Rev 101:1897–1930
49. Shi Z, Lipkowski J (1994) J Electroanal Chem 364:289–294, 365:303–309
50. Ibach H (1997) Surf Sci Rep 29:193–263
51. Leiva EPM, Del Pópolo MG (2000) Schmicler. Chem Phys Lett 320:393–397
52. Wilms M, Broekmann P, Stuhlmann C, Wendelt K (1998) Surf Sci 416:121–140
53. Gregory JE, Gao X, Weaver MJ (1994) J Electroanal Chem 375:357–366
54. Lecoeur J, Bellier JB (1985) Electrochim Acta 30:1027–1033
55. Santos E, Schmickler W (2004) Chem Phys Lett 400:26–29
56. Shi Z, Lipkowski J (1994) J Electroanal Chem 366:317–326
57. Hotlos J, Magnussen OM, Behm RJ (1995) Surf Sci 335:129–144
58. Shi Z, Wu S, Lipkowski J (1995) J Electroanal Chem 384:171–177
59. Shi Z, Wu S, Lipkowski J (1995) Electrochim Acta 40:9–15
60. Baldauf M, Kolb DM (1993) Electrochim Acta 38:2145–2153
61. Kibler LA, Kleinert M, Randler R, Kolb DM (1999) Surf Sci 443:19–30
62. Quayum ME, Ye S, Uosaki K (2002) J Electroanal Chem 520:126–132
63. Stafford GR, Beauchamp CR (2008) J Electrochem Soc 155:D408–D413
64. Woodruff DP, Vlieg E (2002) Surface alloys and alloy surfaces. In: Woodruff DP (ed) The chemical physics of solid surfaces. vol 10. chapt 8. Elsevier, Amsterdam
65. Green MP, Hanson KJ (1991) Surf Sci 259:L743–L749
66. Overbury SH, Mullins DR, Paffett MT, Koel BE (1991) Surf Sci 254:45–57
67. Ku Y, Overbury SH (1992) Surf Sci 273:342–352
68. Overbury SH, Ku Y (1992) Phys Rev B 46:7868–7872
69. Li Y, Koel BE (1995) Surf Sci 330:193–206
70. Umezawa K, Nakanishi S, Yumura T, Gibson WM, Watanabe M, Kido Y, Yamamoto S, Aoki Y, Naramoto H (1997) Phys Rev B 56:10585–10589
71. Brown D, Quinn PD, Woodruff DP, Bailey P, Noakes TCQ (2000) Phys Rev B 61:7706–7715
72. Quinn PD, Bittencourt C, Woodruff DP (2002) Phys Rev B 65:23404-1-4
73. Pauling L (1960) The nature of the chemical bond and the structures of molecules and crystals. 3rd edn. Cornell University Press, New York
74. Woodruff DP, Robinson J (2003) Appl Surf Sci 219:1–10
75. Harrison MJ, Woodruff DP, Robinson J (2004) Surf Sci 572:309–317
76. Seo M (2018) Corrs Rev 36:17–33
77. Seo M, Fushimi K, Aoki Y, Habazaki H, Inaba M, Yokomizo M, Hayakawa T, Nakayama T (2012) J Electroanal Chem 671:7–15
78. Seo M, Habazaki H, Inaba M, Yokomizo M, Wakabayashi T, Nakayama T (2014) J Electrochem Soc 161:H195–H202

Chapter 5
Controversy of Thermodynamics Associated with Surface Stress and Surface Tension of Solid Electrode

Abstract The contradictory discussions made so far in the electrochemistry community for the thermodynamics associated with surface stress and surface tension of a solid electrode have been reviewed. The reviewed subjects include the controversial arguments for the validity of the Gibbs–Duhem equation derived from the excess internal energy for a solid electrode and for the incompatibility of the Shuttleworth equation with Hermann's mathematical structure of thermodynamics. In addition, the review is concerned with the thermodynamic issues for the application of the Shuttleworth equation to a solid electrode and for the surface stress measurement by a cantilever bending method.

Keywords Homogeneous thermodynamic function · Gibbs–Duhem equation · Shuttleworth equation · Hermann's mathematical structure of thermodynamics · Gokhshtein equation

5.1 Introduction

There have been controversial arguments in the electrochemistry community for the thermodynamics associated with surface stress and surface tension of a solid electrode. The Gibbs–Duhem equation for a solid electrode derived by several researchers is not consistent with that derived from the excess internal energy of a solid electrode which is a homogeneous function of the first order with respect to all variables. The contradictory discussions have been made for the validity of the Gibbs–Duhem equation for a solid electrode in connection with the difference between surface stress and surface tension. In addition, the derivations of the Shuttleworth equation, which links surface stress to surface tension, have been argued from the viewpoint of incompatibility with Hermann's mathematical structure of thermodynamics. The controversial discussions still continue up to now.

Reviewing the controversial arguments on the above subjects is essentially necessary to find a clue for solving the thermodynamic issues. Moreover, there have been the thermodynamic issues arose for the application of the Shuttleworth equation

M. Seo, *Electro-Chemo-Mechanical Properties of Solid Electrode Surfaces*,
https://doi.org/10.1007/978-981-15-7277-7_5

to a solid electrode and for the surface stress measurement by a cantilever bending method. The issues result from the lack of thermodynamic constraint such as constant elastic strain or electrode potential. We review the above controversial arguments.

5.2 On Homogeneous Nature of the Thermodynamic Functions of Solid Electrode

Láng and Heusler [1–3] have claimed that the excess internal energy of solid electrode U^{σ} is not a homogeneous function of the first order with respect to all variables. The excess internal energy of the electrode surface can be expressed as functions of surface entropy S^{σ}, surface electric charge Q^{σ}, mole numbers of all surface components n_1^{σ}, …, n_m^{σ}, and surface area A:

$$U^{\sigma} = U^{\sigma}(S^{\sigma}, Q^{\sigma}, n_1^{\sigma}, \ldots, n_m^{\sigma}, A). \tag{5.1}$$

If U^{σ} is a homogeneous function of the first order, the following relationship holds for any real number $\lambda > 0$:

$$\begin{aligned} &U^{\sigma}(\lambda S^{\sigma}, \lambda Q^{\sigma}, \lambda n_1^{\sigma}, \ldots, \lambda n_m^{\sigma}, \lambda A) \\ &= \lambda U^{\sigma}(S^{\sigma}, Q^{\sigma}, n_1^{\sigma}, n_m^{\sigma}, A). \end{aligned} \tag{5.2}$$

Its differentiation with respect to λ provides

$$\begin{aligned} &\frac{\partial U^{\sigma}\left(\lambda S^{\sigma}, \lambda Q^{\sigma}, \lambda n_1^{\sigma}, \ldots., \lambda n_m^{\sigma}, \lambda A\right)}{\partial(\lambda S^{\sigma})}\frac{\partial(\lambda S^{\sigma})}{\partial\lambda} \\ &+ \frac{\partial U^{\sigma}\left(\lambda S^{\sigma}, \lambda Q^{\sigma}, \lambda n_1^{\sigma}, \ldots., \lambda n_m^{\sigma}, \lambda A\right)}{\partial(\lambda Q^{\sigma})}\frac{\partial(\lambda Q^{\sigma})}{\partial\lambda} \\ &+ \frac{\partial U^{\sigma}\left(\lambda S^{\sigma}, \lambda Q^{\sigma}, \lambda n_1^{\sigma}, \ldots., \lambda n_m^{\sigma}, \lambda A\right)}{\partial\left(\lambda n_1^{\sigma}\right)}\frac{\partial\left(\lambda n_1^{\sigma}\right)}{\partial\lambda} \\ &+ \ldots + \frac{\partial U^{\sigma}\left(\lambda S^{\sigma}, \lambda Q^{\sigma}, \lambda n_1^{\sigma}, \ldots., \lambda n_m^{\sigma}, \lambda A\right)}{\partial\left(\lambda n_m^{\sigma}\right)}\frac{\partial\left(\lambda n_m^{\sigma}\right)}{\partial\lambda} \\ &+ \frac{\partial U^{\sigma}\left(\lambda S^{\sigma}, \lambda Q^{\sigma}, \lambda n_1^{\sigma}, \ldots., \lambda n_m^{\sigma}, \lambda A\right)}{\partial(\lambda A)}\frac{\partial(\lambda A)}{\partial\lambda} = U^{\sigma}\left(S^{\sigma}, Q^{\sigma}, n_1^{\sigma}, \ldots, n_m^{\sigma}, A\right). \end{aligned} \tag{5.3}$$

In the case of $\lambda = 1$, Eq. (5.3) takes the form:

$$\frac{\partial U^{\sigma}}{\partial S^{\sigma}}S^{\sigma} + \frac{\partial U^{\sigma}}{\partial Q^{\sigma}}Q^{\sigma} + \frac{\partial U^{\sigma}}{\partial n_1^{\sigma}}n_1^{\sigma} + \ldots + \frac{\partial U^{\sigma}}{\partial n_m^{\sigma}}n_m^{\sigma} + \frac{\partial U^{\sigma}}{\partial A}A = U^{\sigma}. \tag{5.4}$$

Equation (5.4) is an outcome of the Euler's theorem on homogeneous first-order form.

The exact differential of $U^\sigma(S^\sigma, Q^\sigma, n_1^\sigma, \ldots, n_m^\sigma, A)$ is

$$\begin{aligned} \mathrm{d}U^\sigma &= \frac{\partial U^\sigma}{\partial S^\sigma}\mathrm{d}S^\sigma + \frac{\partial U^\sigma}{\partial Q^\sigma}\mathrm{d}Q^\sigma + \frac{\partial U^\sigma}{\partial n_1^\sigma}\mathrm{d}n_1^\sigma \\ &+ \ldots + \frac{\partial U^\sigma}{\partial n_m^\sigma}\mathrm{d}n_m^\sigma + \frac{\partial U^\sigma}{\partial A}\mathrm{d}A. \end{aligned} \tag{5.5}$$

Since $\frac{\partial U^\sigma}{\partial S^\sigma} = T$, $\frac{\partial U^\sigma}{\partial Q^\sigma} = E$, $\frac{\partial U^\sigma}{\partial n_i^\sigma} = \mu_i$, and $\frac{\partial U^\sigma}{\partial A} = \gamma$ (surface tension), Eqs. (5.4) and (5.5) are expressed by

$$U^\sigma = TS^\sigma + EQ^\sigma + \sum_{i=1}^{i=m} \mu_i n_i^\sigma + \gamma A, \tag{5.6}$$

and

$$\mathrm{d}U^\sigma = T\mathrm{d}S^\sigma + E\mathrm{d}Q^\sigma + \sum_{i=1}^{i=m} \mu_i \mathrm{d}n_i^\sigma + \gamma \mathrm{d}A. \tag{5.7}$$

The Gibbs–Duhem equation for a solid electrode is derived mathematically from Eqs. (5.6) and (5.7):

$$S^\sigma \mathrm{d}T + Q^\sigma \mathrm{d}E + \sum_{i=1}^{i=m} n_i^\sigma \mathrm{d}\mu_i + A\mathrm{d}\gamma = 0, \tag{5.8}$$

or

$$s^\sigma \mathrm{d}T + q\mathrm{d}E + \sum_{i=1}^{i=m} \Gamma_i \mathrm{d}\mu_i + \mathrm{d}\gamma = 0, \tag{5.9}$$

where $s^\sigma = \frac{S^\sigma}{A}$, $q = \frac{Q^\sigma}{A}$, and $\Gamma_i = \frac{n_i^\sigma}{A}$. Equation (5.8) or Eq. (5.9) derived from a homogenous function of the first order of U^σ is not consistent with the general form of the Gibbs–Duhem equation for a solid electrode derived by Couchman and Davidson [4] and others [5, 6]:

$$s^\sigma \mathrm{d}T + q\mathrm{d}E + \sum_{i=1}^{i=m} \Gamma_i \mathrm{d}\mu_i + \mathrm{d}\gamma + (\gamma - g)\mathrm{d}\varepsilon = 0, \tag{5.10}$$

where g is surface stress and $\mathrm{d}\varepsilon = \frac{\mathrm{d}A}{A}$ is the change in elastic strain. Equation (5.10) is the form derived for an isotropic solid electrode in place of an anisotropic solid electrode (see Eq. (1.110) in Sect. 1.8 of Chap. 1).

Láng and Heusler [1, 2] argued that Eq. (5.10) is attributed to inserting g into Eq. (5.7) in place of γ, while keeping γ in Eq. (5.6), and they concluded that the use of two different parameters γ and g as the same variable (for the intensive parameter conjugate to A) in the U^σ function cannot be valid and the U^σ function used for derivation of Eq. (5.10) is not a homogeneous function of the first order. In addition, they claimed that Eq. (5.10) is inconsistent with classical thermodynamics since Eq. (5.10) contains the differential of extensive parameter $\mathrm{d}\varepsilon = \frac{\mathrm{d}A}{A}$ despite the Gibbs–Duhem equation representing the relationship between the intensive parameters in the different forms [7]. Gutman [8–10] also claimed that Eq. (5.10) is inconsistent with classical thermodynamics and concluded that Eqs. (5.8) or (5.9) is valid as the Gibbs–Duhem equation for a solid electrode as well as a liquid electrode (such as Hg). However, a distinction between γ and g cannot be made from Eqs. (5.8) or (5.9) which is a homogeneous function independent of whether the surface deformation is plastic or elastic.

Eriksson [11] considered a solid/gas interface consisting of n_1^σ mol of the component from the solid phase and of n_2^σ mol of the component from the gas phase. Since n_1^σ keeps constant (i.e., $\mathrm{d}n_1^\sigma = 0$) under an elastic deformation of the interface, Eriksson wrote the differential of the surface excess of the Helmholtz energy $\mathrm{d}F^\sigma$ as follows:

$$\mathrm{d}F^\sigma = -S^\sigma \mathrm{d}T + g\mathrm{d}A + \mu_2 \mathrm{d}n_2^\sigma. \tag{5.11}$$

Eriksson [11] described that for these kinds of the interface, proportional variations of F^σ, A, n_1^σ, and n_2^σ at constant values of the intensity parameters are physically excluded and thus, there is no integrated relation corresponding directly to Eq. (5.11). This suggests that F^σ cannot be regarded as a homogeneous function of the first order in the case of an elastic deformation.

Guidelli [12] showed that U^σ is not a homogeneous function of the first order for a purely elastic deformation of a solid electrode surface since the constancy ($\mathrm{d}n_1^\sigma = 0$) of the mole number of the surface solid atoms n_1^σ leads to the following inequality:

$$\begin{aligned} &U^\sigma(\lambda S^\sigma, \lambda Q^\sigma, n_1^\sigma, \lambda n_2^\sigma, \ldots, \lambda n_m^\sigma, \lambda A) \\ &\neq \lambda U^\sigma(S^\sigma, Q^\sigma, n_1^\sigma, n_2^\sigma, \ldots, n_m^\sigma, A). \end{aligned} \tag{5.12}$$

In addition, Guidelli [12] recognized that the general form Eq. (5.10) of the Gibbs–Duhem equation for a solid electrode is definitely roughly approximate, but in practice, its application to a solid electrode is usually justified. However, it is clear that thermodynamic equations cannot be roughly approximate.

Although surface area A in Eq. (5.7) is defined as the only state variable describing capillary effects, A may be divided into two terms of A_{p} (related to plastic deformation) and A_{e}(related to elastic deformation) for an isotropic solid electrode [13]:

$$dU^{\sigma} = TdS^{\sigma} + EdQ^{\sigma} + \sum_{i=1}^{i=m} \mu_i dn_i^{\sigma} + \gamma dA_p + gdA_e, \tag{5.13}$$

where $dA = dA_{\mathrm{p}} + dA_{\mathrm{e}}$. In the case of purely plastic deformation ($dA_{\mathrm{e}} = 0$), Eq. (5.13) is equivalent to Eq. (5.7), while in the case of purely elastic deformation ($dA_{\mathrm{p}} = 0$), Eq. (5.13) is equivalent to Eq. (1.120) in Sect. 1.8 of Chap. 1. Kramer and Weissmüller [14] argued that surface area A is a good state variable for the description of fluids, but for solids, A alone is not a good state variable, unless the nature of the change of state is carefully specified. Nevertheless, it still remains unclear whether both A_{p} and A_{e} in the right-hand side of Eq. (5.13) are good state variables or not. If $d\gamma$ is an exact differential of the function of two independent variables E and ε, the Gokhshtein equation (see Eq. (1.119) in Sect. 1.8 of Chap. 1) can be derived from Eq. (5.10). The Gokhshtein equation has been verified [13, 15], which may support the validity of Eq. (5.10).

5.3 Incompatibility of Shuttleworth Equation with Hermann's Mathematical Structure of Thermodynamics

The Shuttleworth equation [16] links surface stress to surface tension for a solid surface (see Eqs. (1.69) and (1.71) in Sect. 1.6 of Chap. 1). Bottomley et al. [17] argued that the derivations of the Shuttleworth equation independently achieved by some researchers [18–21] are inconsistent with Hermann's mathematical structure of thermodynamics [22]. The differential dF^s of the surface contribution to the free energy for the derivation of the Shuttleworth equation may be given by

$$dF^s = d(\gamma A) = Ad\gamma + \gamma dA, \tag{5.14}$$

where γ is the surface tension and A is the surface area. Based on the contact manifold mathematical structure of thermodynamics [17], Hermann's treatment leads to the general equation:

$$dz = \sum_{i=1}^{n} x_i dy_i, \tag{5.15}$$

where z is a thermodynamic function of state, which consists of n pairs of thermodynamically conjugate state variables, the i-th pair of state variables being denoted by x_i and y_i, and the system is represented in a space with exact dimensions of $2n + 1$ ($2n$ state variables plus the state function).

An elementary example of Eq. (5.15) is the case where z is U (internal energy), n is 2, x_1 is $-P$ ($-$ pressure), y_1 is V (volume), x_2 is T (temperature), and y_2 is S (entropy), which is expressed by

$$dU = -PdV + TdS. \tag{5.16}$$

In Eq. (5.16), V is conjugate to $-P$, and S is conjugate to T, implying that the conjugate state variable to any given state variable is unique. By the way, Eq. (5.16) does not contain a term of the form $x_i dy_i + y_i dx_i$, in which x_i and y_i correspond to A and γ, respectively, in Eq. (5.14). A term such as $x_i dy_i + y_i dx_i$ in Eq. (5.14) reduces the exact number of dimensions by two, because of counting twice the state variables x_i and y_i in the term of $x_i dy_i + y_i dx_i$, and thus, the exact number of dimensions required to represent the system is $(2n + 1 - 2)$, i.e., $2n - 1$. In Eq. (5.14), i takes one specific value since the exact number of dimensions is three for $n = 2$, implying that $dF^s = Ad\gamma + \gamma dA$ does not match Hermann's formulation.

Therefore, Bottomley et al. [17] claimed that the derivation of the Shuttleworth equation is not consistent with Hermann's formulation of the mathematical structure of thermodynamics. Furthermore, they have stated that if Hermann's mathematical structure of thermodynamics is true, and if Shuttleworth's equation is a thermodynamic equation, then Shuttleworth's equation is false. Marichev [23] supported the argument of Bottomley et al. [17] with a supplement message that the consistency with Hermann's analysis is necessary, but not sufficient to demonstrate the consistency of thermodynamic equations since if one or more pairs of conjugate state variables changed sign (e.g., from +PV to −PV), the function under consideration would change or completely lose its physical sense, even if remaining formally consistent with the requirements of Hermann's analysis.

Bottomley et al. [24] mentioned later that the consistency with Hermann's theory is a necessary condition. Furthermore, Marichev [23] remarked that the Shuttleworth equation has not been proven experimentally although the Gokhshtein equation is confirmed experimentally [13, 15]. On the other hand, Eriksson and Rusanov [25], and Ibach [26] refuted the argument of Bottomley et al. [17] and asserted the validity of the Shuttleworth equation from different derivations achieved by them. Moreover, Ibach [26] mentioned that Hermann's formal theory concerns the properties of functions defined on the Euclidean space of (independent) Cartesian coordinates, while F^s, A, and γ do not span a Euclidean space since one variable (F^s) is the product of other two variables (A and γ). The corresponding equation for the free energy of bulk systems F would be $F = fV$ (V: the volume and f: the volume specific free energy), where f and V are not conjugate variables as P and V are.

In reply to their refutations, Bottomley et al. [27] argued again that the term of $x_i dy_i + y_i dx_i$ still involves in their different derivations, and the derived Shuttleworth equation is not consistent with Hermann's formulation of the mathematical structure of thermodynamics. In addition, Bottomley et al. [27] claimed that one may define a three-dimensional space represented with respect to an orthogonal basis, the axes representing the state function F and the state variables A and γ, so that there does not seem to be any problem to represent F^s, A, and γ in a Euclidean space, where these quantities obey the equation $F^s = \gamma A$. Afterward, there seems to be no replies between the above prominent researchers and, the derivation of the Shuttleworth equation remains unsolved problem up to now. The final decision of the above arguments may be achieved by whether the validity of the Shuttleworth equation is experimentally confirmed or not.

5.4 Thermodynamic Issues Associated with Shuttleworth Equation and with Surface Stress Measurement by a Cantilever Bending Method

Thermodynamic issues have arisen for the application of the Shuttleworth equation ($g = \gamma + \frac{\partial \gamma}{\partial \varepsilon}$) to an electrified, isotropic solid electrode in electrolyte solution [15]. The first derivative of the Shuttleworth equation with respective to potential has been often described in the literature [e.g., 28–30] as follows:

$$\frac{\partial g}{\partial E} = -q - \frac{\partial q}{\partial \varepsilon} \tag{5.17}$$

However, Proost [15] pointed out that Eq. (5.17) cannot be obtained by differentiation of the Shuttleworth equation. A correct differentiation [15] of the Shuttleworth equation at constant temperature (T) and chemical potential (μ_i) leads to

$$\frac{\partial g}{\partial E} = \frac{\partial \gamma}{\partial E} + \frac{\partial}{\partial E}\left(\frac{\partial \gamma}{\partial \varepsilon}\right). \tag{5.18}$$

The generalized Lippmann equation for an isotropic solid electrode is derived from Eq. (5.10) at constant T and μ_i:

$$\left(\frac{\partial \gamma}{\partial E}\right)_{T,\mu_i} = -q + (g - \gamma)\left(\frac{\partial \varepsilon}{\partial E}\right)_{T,\mu_i}. \tag{5.19}$$

Inserting Eq. (5.19) and the rearranged Shuttleworth equation ($\frac{\partial \gamma}{\partial \varepsilon} = g - \gamma$) into the first term and second term on the right-hand side of Eq. (5.18), respectively, we obtain

$$\frac{\partial g}{\partial E} = -q + (g - \gamma)\frac{\partial \varepsilon}{\partial E} + \frac{\partial}{\partial E}(g - \gamma). \tag{5.20}$$

After rearrangement, Eq. (5.20) simply retrieves the generalized Lippmann equation of Eq. (5.19) for an isotropic solid electrode in place of Eq. (5.17).

Consequently, the Shuttleworth equation cannot be an appropriate starting point to obtain a formulation for $\frac{\partial g}{\partial E}$ [15]. It is emphasized that Eq. (5.19) for a solid electrode should be used as a starting point [15]. The Gokhshtein equation for an isotropic solid electrode, i.e., $\left(\frac{\partial g}{\partial E}\right)_{\varepsilon} = -q - \left(\frac{\partial q}{\partial \varepsilon}\right)_{E}$ explicitly indicates that the partial derivatives of $\left(\frac{\partial g}{\partial E}\right)$ and $\left(\frac{\partial q}{\partial \varepsilon}\right)$ are thermodynamically constrained to constant elastic strain ε and potential E, respectively, as compared to Eq. (5.17). Furthermore, Proost [15] pointed out that the thermodynamic constraint of constant elastic strain does not keep for the surface stress measurement of a solid electrode as a function of applied potential by a cantilever bending method since the curvature of the cantilever electrode, i.e., the elastic strain changes with potential. According to Proost [15], it is strictly inappropriate to use the equations associated with electrocapillarity of a solid electrode derived under the constant elastic strain in order to interpret the surface stress versus potential curve measured by a cantilever bending method. The average value of $\frac{\Delta g}{\Delta E}$ in some potential range obtained from the surface stress versus potential curve by using the cantilever bending method does not meet the thermodynamic constraint of constant elastic strain.

References

1. Láng G, Heusler KE (1994) J Electroanal Chem 377:1–7
2. Láng G, Heusler KE (1999) J Electroanal Chem 472:168–173
3. Láng GG, Barbero CA (2012) Laser techniques for electrode processes. chap 3. Springer, Berlin
4. Couchman PR, Davidson CR (1977) J Electroanal Chem 85:407–409
5. Linford RG (1978) Chem Rev 78:81–95
6. Schmickler W, Leiva E (1998) J Electroanal Chem 453:61–67
7. Callen HB (1960) Thermodynamics: an introduction to the physical theories of equilibrium thermostatics and irreversible thermodynamics. chap 3. Wiley, New York
8. Gutman EM (1995) J Phys Condens Mater 7:L663–L667
9. Gutman EM (2014) J Solid State Electrochem 18:3217–3237
10. Gutman EM (2016) J Solid State Electrochem 20:2929–2950
11. Eriksson JC (1969) Surf Sci 14:221–246
12. Guidelli R (1999) J Electroanal Chem 472:174–177
13. Valincius G (1999) J Electroanal Chem 478:40–49
14. Kramer D, Weissmüller J (2007) Surf Sci 601:3042–3051
15. Proost J (2005) J Solid State Electrochem 9:660–664
16. Shuttleworth R (1950) Proc Phys Soc London A63:444–457
17. Bottomley DJ, Makkonen L, Kolari K (2009) Surf Sci 603:97–101
18. Ibach H (1997) Surf Sci Rep 29:193–263
19. Müller P, Saúl A (2004) Surf Sci Rep 54:157–258
20. Rusanov AI (1996) Surf Sci Rep 23:173–247
21. Rusanov AI (2005) Surf Sci Rep 58:111–239
22. Hermann R (1973) Geometry, physics, and systems. chap. 6. Marcel Dekker Inc, New York
23. Marichev VA (2009) Surf Sci 603:2345–2346
24. Bottomley DJ, Makkonen L, Kolari K (2009) Surf Sci 603:2347

25. Erikkson JC, Rusanov A (2009) Surf Sci 603:2348–2349
26. Ibach H (2009) Surf Sci 603:2352–2355
27. Bottomley DJ, Makkonen L, Kolari K (2009) Surf Sci 603:2350–2351, 2356–2357
28. Haiss W, Sass JK (1996) J Electroanal Chem 410:119–124
29. Haiss W, Nichols RJ, Sass JK, Charle KP (1998) J Electroanal Chem 452:199–202
30. Seo M, Serizawa Y (2003) J Electrochem Soc 150:E472–E476

Chapter 6
Stresses of Anodic Oxide Films Grown on Metal Electrode

Abstract Stresses are generated in anodic oxide films grown on valve metals during anodic oxidation. The stress generated in anodic oxide film, i.e. the film stress is influenced by many electrochemical and physical factors associated with the formation and growth mechanism of anodic oxide film such as volume change, transport number of mobile species in the films, electrostriction, crystallization, and oxygen evolution. The main factors influencing the film stress during anodic oxidation are separately explained, and the contributions of the respective factors to the film stress are discussed on the basis of the typical experimental results for anodic oxidation of Al and Ti. Furthermore, the stress variation toward compressive direction during cathodic polarization of the anodic oxide film on Ti is explained in terms of the volume expansion due to hydrogen uptake of the film, and the plastic flow driven by the compressive stress in the barrier layer on Al is discussed with relation to the formation of the porous layer.

Keywords Film stress · Pilling–Bedworth ratio · Film growth · Transport number · Electrostriction

6.1 Introduction

A cantilever bending method has been employed for the measurement of film stress as well as surface stress. The measured quantity is the product of film stress σ_f and thickness d_f with a unit of N m^{-1} or J m^{-2} which is equivalent to that of surface stress g, i.e. $\Delta g = \Delta(\sigma_f \cdot d_f)$. The unit of film stress, therefore, is Pa or J m^{-3}. If the change in film thickness is known, an average value $\overline{\sigma}_f$ of film stress can be obtained by dividing Δg by Δd_f. Stresses are generated in anodic oxide films grown on valve metals such as Al and Ti during anodic oxidation. The anodic oxide films on valve metals have been used as electrolytic capacitors in the electronic industry. The stress generated in anodic oxide film, i.e. the film stress is influenced by the electrochemical and physical factors associated with the film formation and growth mechanism, such as volume change, transport number of mobile species, electrostriction, crystallization, and oxygen evolution. The performance of the anodic oxide films as electrolytic

M. Seo, *Electro-Chemo-Mechanical Properties of Solid Electrode Surfaces*,
https://doi.org/10.1007/978-981-15-7277-7_6

capacitors may be linked to the magnitude and sign (tensile or compressive) of the film stress.

The compressive or tensile stress is generated in the anodic oxide film, depending on the factor which is predominant for the stress generation, e.g., if the film growth is controlled by mobile oxygen ion in the film, the compressive stress is generated in the film side near the metal/film interface, while if the film growth is controlled by mobile metallic cation in the film, the tensile stress is generated in the metal side near the metal/film interface. The growth of anodic oxide films on valve metals obeys a high field model in which the film growth is controlled by the high electric field in the film. The high electric field induces the compressive stress in the film due to electrostriction during film growth. Furthermore, in the case where the internal residual stress is present in the substrate metal, the relaxation of the residual stress resulting from the decrease in the substrate thickness due the film growth has to be taken into consideration to evaluate the stress of the film itself. Therefore, the analysis of the stress component for the anodic oxide film is complicated because many factors influencing the film stress are overlapped. It is essentially necessary to separate the main factors influencing the film stress and to analyze the contributions of the respective factors to the film stress for understanding of the generation mechanism of the film stress.

A cathodic polarization of the anodic oxide film induces the changes in film stress. The compositional changes of the anodic oxide film during cathodic polarization are closely linked to the changes in film stress. If the compressive stress in the film exceeds a critical level of elastic deformation, the film is plastically deformed. The plastic flow of the film may lead to the growth of a porous layer. In the present chapter, we discuss the main factors influencing the film stress and the contributions of the respective factors to the stresses of the anodic oxide films on Al and Ti. Moreover, we discuss the changes in film stress due to the alteration of the anodic oxide film on Ti during cathodic polarization and the contribution of plastic flow to the formation of porous layer on Al.

6.2 High Field Model for Growth of Anodic Oxide Film

The stress generation and the alteration of anodic oxide film formed on metal are directly associated with the mechanism of film growth. Here, let us explain briefly a high field model [1, 2] for the growth of anodic oxide films on valve metals such as Al, Ta, and Zr as fundamentals for understanding of many factors influencing the film stress. The high field model for the oxide growth is based on the ion hopping mechanism in which the ions placed at regular or interstitial sites in the oxide lattice can jump to the neighboring vacancy or other interstitial sites.

Figure 6.1 shows schematically the potential energy of mobile ion versus distance coordinate with and without an applied potential [3]. It is assumed that the ion vibrates in simple harmonic mode with a frequency $\nu\left(s^{-1}\right)$ at the minimum point of potential energy corresponding to the lattice plane of the oxide film. In the absence of an

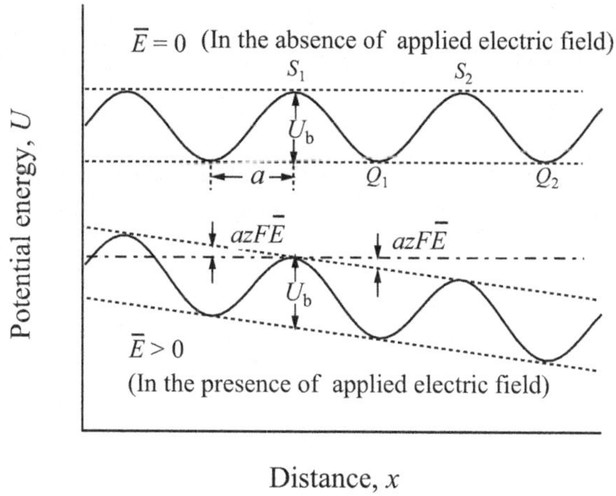

Fig. 6.1 Schematic illustration for the potential energy of mobile ion versus distance coordinate with and without an applied potential [3]. The symbols shown are Q_1, Q_2, ...: the minimum points of the potential energy corresponding to the lattice plane of oxide film; $\bar{E}$: the applied electric field; U_b: the potential barrier for the mobile ion in the absence of $\bar{E}$; S_1, S_2, ...: the top points of the potential barrier; a: the half barrier distance or half-jump distance; z: the charge number of the mobile ion; and F: the Faraday constant. Reprinted from [3], Copyright 1993, with permission from Elsevier

electric field, a probability P_b of the ion which will possess sufficient energy to jump a barrier U_b and to reach the next site is given by

$$P_b = \nu \exp\left(-\frac{U_b}{RT}\right), \tag{6.1}$$

where R is the gas constant, and T is the Kelvin temperature (K). When an electric field $\bar{E}$ is applied, the barrier height for the ion to move with the field reduces from U_b to $U_b - azF\bar{E}$ (a is the half barrier distance or half-jump distance, z is the charge number of the mobile ion, and F is the Faraday constant). On the other hand, the barrier height for the ion to move against the field increases from U_b to $U_b + azF\bar{E}$. If the concentration in mol of the mobile ion per unit volume is denoted by c_m (mol m^{-3}), the amount of the mobile ion n_m (mol m^{-2}) per unit surface area of the lattice plane can be given by

$$n_m = c_m 2a. \tag{6.2}$$

In general, c_m is a function of position x through the oxide film [2]. The transfer rate of the mobile ion $\frac{dn_m}{dt}$ in the presence of the electric field can be written by

$$\frac{dn_{\mathrm{m}}}{dt} = 2a\nu \exp\left(-\frac{U_{\mathrm{b}}}{RT}\right)\left[c_{\mathrm{m}} \exp\left(\frac{azF\bar{E}}{RT}\right) - \left(c_{\mathrm{m}} + 2a\frac{dc_{\mathrm{m}}}{dx}\right)\exp\left(-\frac{azF\bar{E}}{RT}\right)\right], \tag{6.3}$$

where $\frac{dc_{\mathrm{m}}}{dx}$ is the concentration gradient of the mobile ion due to diffusion in the film.

Since the electric field is usually high enough to neglect the effect of diffusion effect ($\frac{dc_{\mathrm{m}}}{dx} = 0$), Eq. (6.3) is reduced to

$$\frac{dn_{\mathrm{m}}}{dt} = 2a\,\nu c_{\mathrm{m}} \exp\left(-\frac{U_{\mathrm{b}}}{RT}\right)\left[\exp\left(\frac{azF\bar{E}}{RT}\right) - \exp\left(-\frac{azF\bar{E}}{RT}\right)\right]. \tag{6.4}$$

Moreover, if the electric field is high enough to prevent the ion movement against the field, Eq. (6.4) can be simplified:

$$\frac{dn_{\mathrm{m}}}{dt} = 2a\nu c_{\mathrm{m}} \exp\left(-\frac{U_{\mathrm{b}}}{RT}\right)\exp\left(\frac{azF\bar{E}}{RT}\right). \tag{6.5}$$

The transfer rate of the mobile ion in Eq. (6.5) can be converted to the anodic current density i_{a} (A m^{-2}) for the film growth:

$$i_{\mathrm{a}} = zF\frac{dn_{\mathrm{m}}}{dt} = 2a\nu c_{\mathrm{m}} zF \exp\left(-\frac{U_{\mathrm{b}}}{RT}\right)\exp\left(\frac{azF\bar{E}}{RT}\right). \tag{6.6}$$

The simplified form of Eq. (6.6) which represents the relationship between the anodic current density and electric field for the growth of the anodic oxide film is expressed by

$$i_{\mathrm{a}} = i_* \exp(\beta\bar{E}), \tag{6.7}$$

$$i_* = 2a\nu c_{\mathrm{m}} zF \exp\left(-\frac{U_{\mathrm{b}}}{RT}\right), \tag{6.8}$$

and

$$\beta = \frac{azF}{RT}. \tag{6.9}$$

In the case where the anodic oxide film is homogeneous, the electric field $\bar{E}$ in the film can be estimated by using the potential drop $\Delta\phi_{\mathrm{f}}$ across the film and the film thickness d_{f}:

$$\bar{E} = \frac{\Delta\phi_{\mathrm{f}}}{d_{\mathrm{f}}}. \tag{6.10}$$

Consequently, Eq. (6.7) is represented by

$$i_a = i_* \exp\left(\beta \frac{\Delta\phi_f}{d_f}\right). \tag{6.11}$$

It is known that Eq. (6.7) holds for the growth of a barrier-type film during anodic oxidation of Al and Ta [4]. The linear relation between $\ln i_a$ and $\bar{E} = \frac{\Delta\phi_f}{d_f}$ has been also confirmed by ellipsometry for anodic oxidation of Ti under a galvanostatic condition of $i_a = 2.0 \times 10^{-7} - 1.0 \times 10^{-5}$ A cm^{-2} in pH 6.9 phosphate solution [5]. Khalil and Leach [6] reported the values of $\bar{E} = 9.1 \times 10^8, 6.2 \times 10^8$, and 5.0×10^8 V m^{-1}, respectively, for the anodic oxide films formed on Al, Ta, and Zr under a galvanic condition of $i_a = 6 \times 10^{-3}$ A cm^{-2} up to a cell voltage of 100 V in ammonium hydrogen tetraborate (NH_4BO_2) solution.

If the anodic oxidation proceeds with a current efficiency of 100% for the film growth, i_a can be converted to the film growth rate $\frac{d(d_f)}{dt}$:

$$\frac{d(d_f)}{dt} = \frac{V_{ox}}{S_r z F} i_a = \frac{V_{ox}}{S_r z F} i_* \exp\left(\beta \frac{\Delta\phi_f}{d_f}\right), \tag{6.12}$$

where V_{ox} and S_r are the molar volume and the surface roughness of the anodic oxide film, respectively. Equation (6.12) is further deformed to

$$\exp\left(-\beta \frac{\Delta\phi_f}{d_f}\right) d(d_f) = K dt, \tag{6.13}$$

where $K = \frac{V_{ox}}{S_r z F} i_*$ is regarded as a constant. The approximate solution of Eq. (6.13) under the condition $d_f \ll \beta\Delta\phi_f$ has been derived by Cabrera and Mott [1]:

$$\frac{d_f^2}{\beta\Delta\phi_f} \exp\left(-\beta \frac{\Delta\phi_f}{d_f}\right) = Kt + \text{constant}, \tag{6.14}$$

where the value of constant is zero since d_f is defined as zero when t is zero. Taking logarithms of both sides of Eq. (6.14), we obtain

$$2 \ln d_f - \ln \beta\Delta\phi_f - \beta \frac{\Delta\phi_f}{d_f} = \ln K + \ln t. \tag{6.15}$$

Under the condition of $\beta \frac{\Delta\phi_f}{d_f} \gg 2 \ln d_f$, the following relationship [1, 7] is derived from Eq. (6.14):

$$\frac{\beta\Delta\phi_f}{d_f} = -\ln \beta\Delta\phi_f - \ln K - \ln t. \tag{6.16}$$

Equation (6.16) shows that at constant $\Delta\phi_f$ and T, d_f is inversely proportional to $\ln t$, which is named "the inverse logarithmic law of oxide growth". Taking logarithms of both sides of Eq. (6.11) and then substituting Eq. (6.16) into Eq. (6.11), we obtain

$$\ln i_a = -\ln \beta \Delta\phi_f - \ln \frac{V_{ox}}{S_r z F} - \ln t. \quad (6.17)$$

Equation (6.17) indicates that the relationship between $\ln i_a$ and $\ln t$ has a slope of −1 in the time domain where the anodic oxidation of valve metals at constant $\Delta\phi_f$ (i.e., during potentiostatic polarization or under constant cell voltage) obeys the inverse logarithm law.

6.3 Pilling–Bedworth Ratio

The Pilling–Bedworth ratio α_{PB} has been often used to predict the sign (either minus, i.e. compressive or plus, i.e., tensile) of stress generated during growth of an oxide film on a metal [8]. α_{PB} is defined by

$$\alpha_{PB} = \frac{V_{ox}}{x V_m} = \frac{M_{ox} \rho_m}{x M_m \rho_{ox}}, \quad (6.18)$$

where x is the stoichiometric number of metal component in the oxide (Me_xO_y), V_{ox} is the molar volume of the oxide, V_m is the molar volume of the metal, M_{ox} is the molecular weight of the oxide, M_m is the atomic weight of the metal, ρ_{ox} is the density (g cm^{-3}) of the oxide, and ρ_m is the density of the metal. The value of α_{PB}, therefore, represents the relative volume change per the metal component due to oxidation of metal. In the case of $\alpha_{PB} > 1$, the sign of the stress generated by oxidation is minus (i.e. compressive) because of the volume expansion. On the other hand, in the case of $\alpha_{PB} < 1$, the sign of the stress generated is plus (i.e. tensile) because of the volume shrinkage.

Figure 6.2 shows schematically that if the one side of a metal strip is oxidized, the bending of the strip in the oxidation side is convex due to the generation of compressive stress for $\alpha_{PB} > 1$, while it is concave due to the generation of tensile stress for $\alpha_{PB} < 1$. The values of α_{PB} for various metal/metal oxide systems are listed in Table 6.1. The numerical values of α_{PB} with and without mark* in Table 6.1 were calculated from Eq. (6.18) by employing the values of ρ_{ox} for crystalline metal oxides in refs [9] * and [10], respectively. It is remarked that the values of α_{PB} for amorphous or hydrous metal oxides become large as compared to those for crystalline metal oxides in Table 6.1 because amorphous or hydrous metal oxides have low values of ρ_{ox}.

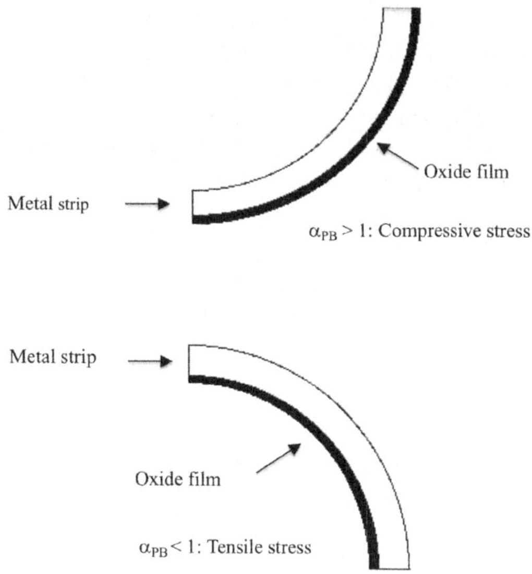

Fig. 6.2 Schematic representation of the bending of a metal strip when the one side of the strip is oxidized to form a thin oxide film. The oxide film with a value of $\alpha_{PB} > 1$ or $\alpha_{PB} < 1$ is subjected to compressive stress or tensile stress, respectively, and the resultant bending of the strip is convex or concave

Table 6.1 Values of α_{PB} calculated from Eq. (6.18) for various metal/metal oxide systems

Metal/Metal Oxide	α_{PB}	Metal/Metal Oxide	α_{PB}
Li/Li_2O	0.44*	Zn/ZnO	1.59
Be/BeO	1.70	Zr/ZrO_2	1.58
Mg/MgO	0.79	Nb/Nb_2O_5	2.47
Al/Al_2O_3	1.29	Mo/MoO_3	3.29
Si/SiO_2	2.15	Ag/Ag_2O	1.58*
Ca/CaO	0.63	Cd/CdO	1.42
Ti/TiO_2	1.77	α-Sn/SnO	1.02
V/V_2O_5	3.32	β-Sn/SnO	1.29
Cr/Cr_2O_3	2.02	α-Sn/SnO_2	1.05
Mn/MnO_2	2.34	β-Sn/SnO_2	1.33
Fe/FeO	1.77	Ba/BaO	0.68
Fe/Fe_3O_4	2.08	Hf/HfO_2	1.59
Fe/Fe_2O_3	2.14	Ta/Ta_2O_5	2.33
Co/Co_2O_3	2.42	W/WO_3	3.75
Ni/NiO	1.52	Pb/PbO	1.29*
Cu/Cu_2O	1.68	Pb/PbO_2	1.40
Cu/CuO	1.75	Bi/Bi_2O_3	1.23

The numerical values of α_{PB} with and without mark (*) in Table 6.1 were calculated from Eq. (6.18) by employing the values of ρ_{ox} for crystalline metal oxides in refs [9] * and [10], respectively

Although most of the metal/metal oxide systems have $\alpha_{PB} > 1$, the metal/metal oxide systems such as Li/Li_2O, Mg/MgO, Ca/CaO, and Ba/BaO have $\alpha_{PB} < 1$, predicting the generation of tensile stress due to the oxide film growth. The tensile strength (96 MPa) of MgO is less than one order of magnitude as much as its compressive strength (1.4 GPa) [10]. The MgO film on Mg may be non-protective because of the film fracture or breakdown due to the tensile stress generated during the oxide film growth. However, in many cases, the real sign of stresses measured during the oxide film growth is likely opposite to that predicted from α_{PB}. In addition to α_{PB}, the transport (or transference) number of mobile ion in the oxide film during film growth is one of main factors that control the sign and magnitude of stresses as explained in the next section.

6.4 Transport Number of Mobile Ion in Anodic Oxide Film and Stress Generation

For an anodic oxidation of metal, water molecule is dissociated to $2H^+$ and O^{2-} at the oxide film/solution interface, while metal atom Me in the metal side at the metal/oxide film interface is ionized to metal ion Me^{z+} by an oxidation reaction ($Me \rightarrow Me^{z+} + ze^-$). The oxygen ion O^{2-} generated at the oxide film/solution interface is transported inward via vacancy site of O^{2-}ion under a high electric field in the film to form a new oxide by reacting with Me^{z+} ion at the metal/oxide film interface. On the other hand, Me^{z+} ion at the metal/oxide film interface is transported outward via vacancy or interstitial site in the film to form a new oxide by reacting with O^{2-} ion at the oxide film/solution interface. In the case where both O^{2-} and Me^{z+} ions are mobile for the film growth, the anodic current density i_a consists of partial current densities i_o due to the transport of O^{2-} ion and i_m due to the transport of Me^{z+} ion:

$$i_a = i_o + i_m \tag{6.19}$$

The transport number of each mobile ion is defined with each partial current density divided by total current density:

$$t_o = \frac{i_o}{i_a} = \frac{i_o}{i_o + i_m}, \tag{6.20}$$

$$t_m = \frac{i_m}{i_a} = \frac{i_m}{i_o + i_m}, \tag{6.21}$$

and

$$t_o + t_m = 1, \tag{6.22}$$

where t_o and t_m are the transport numbers of O^{2-} and Me^{z+} ions, respectively.

For practical determination of the transport number of mobile ion during anodic oxidation of valve metals, a thin surface layer is tagged with a completely immobile atom (served as Kirkendall marker), whose position is then pursued during the subsequent anodic oxidation. After the anodic oxidation, the transport number of the mobile ion can be determined from the position of the marker atom in the oxide film since the part of the oxide film above the marker position is due to the transport of Me^{z+} ion, while the part of the oxide film underneath the marker position is due to the transport of O^{2-} ion. Radiotracer techniques of inert gas atom such as Xe^{125} and Rn^{222} [6, 11] and nuclear micro-analysis of O^{18}/O^{16} [12–15] have been used to investigate the transport of the mobile ion in the anodic oxide films on valve metals. Rutherford backscattering spectroscopy (RBS) [16] was also used to determine the position of ion-implanted marker atoms (Xe, Ar, Kr, etc.) during anodic oxidation of Al. If Me^{z+} ion does not dissolve into solution through the oxide film during anodic oxidation, t_m can be determined from

$$t_{m} = \frac{d_{mr,f}}{d_{t,f}}, \tag{6.23}$$

where $d_{mr,f}$ is the thickness of the oxide film above the marker position, and $d_{t,f}$ is the total thickness of the oxide film. On the other hand, in the case where some amount of Me^{z+} ion dissolves into solution through the oxide film during anodic oxidation, t_m is modified to

$$t_{m} = \frac{d_{mr,f} + \xi d_{t,f}}{(1+\xi)d_{t,f}}, \tag{6.24}$$

where ξ is the fraction of $d_{t,f}$ corresponding to the dissolved amount of Me^{z+} ion.

For the prediction of the sign of film stress from the Pilling–Bedworth ratio in Eq. (6.18), it is implicitly assumed that O^{2-} ion generated at the oxide film/solution interface is solely transported inward to form a new oxide only at the metal/oxide film interface during film growth, i.e. t_o is unity. Since the metal/oxide film interface is mechanically constrained, a volume change due to the oxide formation at the metal/oxide film interface induces a stress. As shown schematically in Fig. 6.3a, in the case of $\alpha_{PB} > 1$, the volume expands due to the oxide formation at the metal/oxide film interface, which induces the compressive stress. By contrast, if Me^{z+} ion at the metal/oxide film interface is solely transported outward, the oxide film/solution interface, i.e. t_m is unity, new oxide forms only at the oxide film/solution interface. As shown schematically in Fig. 6.3b, the vacancy of metal atom in the metal side at the metal/oxide film interface is created due to the oxidation of Me atom, followed by the transfer of Me^{z+} ion toward the oxide film/solution interface. The vacancy of metal atom provides a free space in the metal side to induce tensile stress as far as the metal vacancy does not sink into the metal substrate. The creation of the metal vacancy in the metal side at the metal/oxide film interface induces tensile stress.

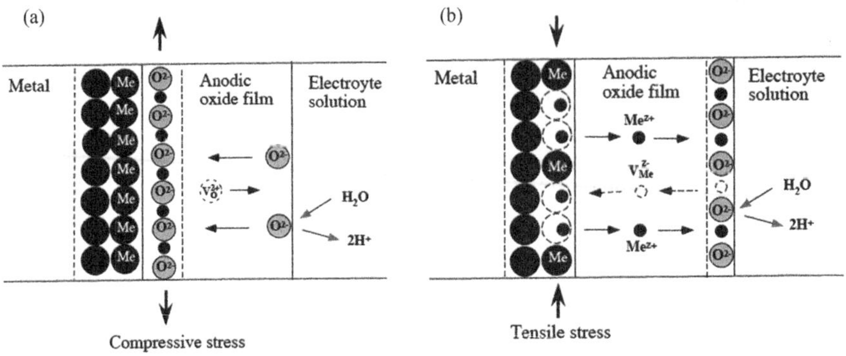

Fig. 6.3 Schematic diagram of the stress generation at the metal/oxide film interface: **a** the compressive stress generation in the case where O^{2-} ion at the oxide film/solution interface is solely transported inward to form a new oxide only at the metal/oxide film interface during film growth, i.e., t_o is unity, and **b** the tensile stress generation in the case where Me^{z+} ion at the metal/oxide film interface is solely transported outward the oxide film/solution interface to form a new oxide only at the oxide film/solution interface, i.e., t_m is unity. The symbols V_O^{2+} and V_{Me}^{z-} are the oxygen ion vacancy with a charge number of 2+ and the metal ion vacancy with a charge number of z−, respectively. The transport directions of V_O^{2+} and V_{Me}^{z-} are opposite to those of O^{2-} and Me^{z+} ions

On the other hand, the oxide formation at the oxide film/solution interface does not induce any stresses since the oxide film/solution interface is not mechanically constrained. Consequently, the compressive stress is not always generated at $\alpha_{PB} > 1$ in the case of $t_o < 1$.

6.5 Criterion for Stress Generation by Nelson and Oriani

Nelson and Oriani [17] have proposed that the sign (compressive or tensile) of stress generated during anodic oxidation is determined by the ratio f_v of the volume occupied by new oxide at the metal/oxide interface to the available free space caused by the ionization of metal atom at the same interface. The ratio f_v is given by

$$f_v = \frac{t_o V_{ox}}{x V_m} = t_o \alpha_{PB}. \tag{6.25}$$

All symbols in Eq. (6.25) are same as those in Eqs. (6.18) and (6.20). The denominator $x\,V_m$ in Eq. (6.25) corresponds to the available free space caused by the ionization of metal atom at the metal/oxide film interface, while the numerator $t_o\,V_{ox}$ corresponds to the volume occupied by new oxide at the same interface. No stresses are generated in the case of f_v being unity, i.e., $t_o = (\alpha_{PB})^{-1}$. The critical value of t_o being equal to reciprocal of α_{PB} is defined with t_o^c. If t_o is larger than t_o^c, the compressive stress is generated due to the excess volume of the oxide formed at the metal/oxide film interface. By contrast, if t_o is less than t_o^c, the tensile stress is generated due to the free space remaining in the metal side at the metal/oxide film interface. However, it is reminded that the other factors influencing the sign of the film stress such as electrostriction and crystallization are not taken into consideration for Nelson–Oriani's criterion [17].

As explained in the next section, the compressive stress due to the electrostriction is generated in the oxide film grown during anodic oxidation and can be measured from the stress relaxation caused by switching off the galvanostatic current or by returning the applied potential to open-circuit potential. In Table 6.2, the sign of the film stress predicted from the values of α_{PB} and t_o [17] (i.e. Nelson–Oriani's criterion) is compared with the sign of the film stress obtained by subtracting the electrostriction component (see Sect. 6.6 of this chapter) from the net stress changes which were measured for anodic oxidation of Ti, Al, Zr, Ta, W, Nb, and Hf. For the stress measurements of anodic oxide films, the metal strips coated on one side with a flexible lacquer or enamel were mostly used in early studies up to the 1990's. On the other hand, the thin metal films evaporated or sputter-deposited on a glass plate or Si (100) wafer have been often used from the 2000's. In Table 6.2, the data of the film stress obtained in the 2000's are added to the original table in Ref. [17]. It is reminded that the numbers with superscript of b in the seventh column of Table 6.2 represent the references obtained in the 2000's. Although the observed stress of the anodic oxide film on Ti is tensile (T) in the original table [17], the observed film stress for anodic oxidation of the sputter-deposited or evaporated Ti thin films on glass plate or Si wafer is compressive (C) [19] or compressive and tensile (C and T) [20, 21].

As seen from Table 6.2, the sign of the observed film stress (tensile) is opposite to that of the predicted film stress (compression) for anodic oxidation of Ta, W, and Nb. The observed tensile stress may be associated with crystallization of an

Table 6.2 Properties (α_{PB}, t_o^c and t_o) and stress information (T for tensile and C for compression) of anodic oxide films formed on selected metals [17] Modified from [17], Copyright 1993, with permission from Elsevier

Metal	α_{PB}	t_o^c	t_o	Predicted stress	Observed stress	References
Al	1.3–1.7	0.59–0.77	0.28–0.67[a]	T and C	T and C	[18[b], 22, 23, 25, 60]
Ti	1.7–2.4	0.42–0.59	0.61–0.65	C or T	C and/or T	[6, 19[b]–21[b], 61, 62]
Zr	1.5	0.67	0.88–1.0	C	C	[5, 11, 23, 24[b], 63]
Ta	2.5–2.6	0.38–0.40	0.66–0.74	C	T	[6, 11, 22, 23]
W	3.30	0.30	0.63-0.70	C	T	[11, 22]
Nb	2.4-2.8	0.36-0.42	0.67-0.78	C	T	[11, 22, 23]
Hf	1.6	0.63	0.95	C	C	[11, 23]

The data of the film stress obtained in the 2000's are added to Table 6.2
[a] Varies with current density
Refs [b] obtained in the 2000's

amorphous oxide film formed during anodic oxidation under a high galvanostatic current density [22, 23] since the increase in density of anodic oxide film due to crystallization induces tensile stress because of the volume contraction of the film. On the other hand, the sign of the film stress observed during anodic oxidation of Zr and Hf satisfies Nelson–Oriani's criterion [17]. Particularly, the stress measurement during anodic oxidation of a sputter-deposited Zr thin film in sulfuric acid solution [24] confirmed that the stress generated in anodic oxide film on Zr is compressive and satisfies Nelson–Oriani's criterion. In the cases of Al and Ti, the assessment for Nelson–Oriani's criterion [17] is complicated because the signs of both predicted and observed film stresses vary depending on the experimental conditions of anodic oxidation. We explain the stress generation behavior during anodic oxidation of Al and Ti in the following subsections and discuss the matching with Nelson–Oriani's criterion [17].

6.5.1 Stress Generation during Anodic Oxidation of Al

The stress generated during anodic oxidation of Al in ammonium borate and ammonium citrate solutions has been measured as a function of galvanostatic current density [25]. The stress of the anodic oxide film itself obtained by subtracting the compressive stress component due to electrostriction is compressive at low current density and tensile at high current density. The stress transition from compressive to tensile takes place at about 0.5 mA cm^{-2}. The stress transition from compressive to tensile at 0.6 mA cm^{-2} was also observed for the film stress generated during anodic oxidation of Al in 0.1 M H_2SO_4 solution [17]. In general, the anodic oxide film formed on Al at high current density is amorphous, while the film formed at low current density contains crystalline [23]. The values of $\alpha_{PB} = 1.3 - 1.4$ and $1.6 - 1.7$ have been reported for crystalline Al_2O_3 on Al and for glassy Al_2O_3 on Al, respectively [23]. According to the measurement of the transport number of mobile ion in anodic oxide films [11], the transport number of O^{2-} ion t_o decreases from 0.67 to 0.28 with increasing current density from 0.1 to 10 mA cm^{-2} for anodic oxidation of Al in ammonium citrate solution.

If the values of $t_o = 0.4$ and $\alpha_{PB} = 1.7$ are employed for the amorphous film formed on Al at a current density higher than 0.5 mA cm^{-2}, an inequality of $t_o < t_o^c = 0.59$ holds, predicting the tensile film stress from Nelson–Oriani's criterion [17], which is consistent with the sign of the observed film stress. On the other hand, if the values of $t_o = 0.67$ and $\alpha_{PB} = 1.55$ are employed for the crystalline-containing film formed on Al at a current density lower than 0.5 mA cm^{-2}, t_o is slightly larger than $t_o^c = 0.65$, suggesting that the sign of the film stress is minus (compressive), which is consistent with the sign of the observed film stress. However, it seems difficult to predict the sign of the film stress because the predicted sign of the film stress on Al at the low current density is reversed depending on the selected values of t_o and α_{PB}. The recent study of the stress variations during the growth of anodic oxide film on Al up to 25 V at galvanostatic current densities of $2.0 - 12.5$ mA cm^{-2}

in 0.4 M H_3PO_4 solution, followed by dissolution of the oxide film in open-circuit in the same solution [18] indicated that the stress of the oxide film itself after removal of the compressive stress component due to the electrostriction is compressive, while the stress varies to tensile by dissolution of the oxide film. This means that the tensile stress near the metal/oxide interface is masked with the compressive stress in the oxide film. The stress in the oxide film near the oxide/solution interface is not accounted for Nelson–Oriani's criterion [17] since the oxide film/solution interface is regarded as mechanically unconstrained. Consequently, Nelson–Oriani's criterion [17] is not sufficient to explain all the results of the stress generation during anodic oxidation of Al under various experimental conditions.

6.5.2 Stress Generation during Anodic Oxidation of Ti

The stress generation behavior during anodic oxidation of Ti is much complicated as compared to that during anodic oxidation of Al. According to the stress measurement during anodic oxidation of a sputter-deposited Ti thin film (with a thickness of 250 nm) on a glass plate by a potential step up to 10.7 V (RHE) in pH 8.4 borate solution [19] as shown in Fig. 6.4b, the value of $(\sigma_f \cdot d_f)$ at 1 h-anodic oxidation is compressive and decreases toward compressive direction with increasing applied potential E, i.e. with increasing thickness of the anodic oxide film d_f (see Fig. 6.4c). The anodic current density i_a (see Fig. 6.4a) increases with increasing E, but it does not exceed 60 μA cm^{-2} at 10.7 V (RHE). The thickness of the anodic oxide film d_f in Fig. 6.4c was converted from the thickness of the anodic oxide film measured by ellipsometry at 1 h-potentiostatic oxidation of Ti in pH 6.9 phosphate solution [5]. Here, it is reminded that a potential (RHE) in the abscissa of Fig. 6.4 is referred to a reversible hydrogen electrode (pH $\neq$ 0) and can be converted to a potential (SHE), referred to a standard reversible hydrogen electrode (pH $=$ 0) by using the relationship of $E(\text{RHE}) = E(\text{SHE}) + 0.059\text{pH}$. The use of E (RHE) is convenient to compare the thickness of anodic oxide films formed on metals as a function of applied potential in solutions with different pH values.

The average compressive stress of $\sigma_f \approx -500$ MPa for the anodic oxide films formed on Ti in the potential region between 1.7 V and 10.7 V (RHE) is obtained from the linear relation between $(\sigma_f \cdot d_f)$ and d_f. The electrostriction component of about -20 MPa was estimated from the stress relaxation during a cathodic potential scan of 1 mV s^{-1} from the film formation potential to a flat band potential of 0.0 V (RHE) [26] for the anodic oxide film after 1 h-anodic oxidation. As listed in Table 6.2, for anodic oxidation of Ti, $t_o = 0.61 - 0.65$ is close to $t_o^c = 0.42 - 0.59$. According to laser Raman spectroscopic study [27], an anatase type of TiO_2 film grows during anodic oxidation of Ti at potentials higher than 4 V (RHE) in pH 6.9 phosphate solution. If an anatase type of the film is formed, the value of $\alpha_{PB} = 1.96$ is obtained by substituting $V_{ox} = 20.8\,\text{cm}^3\,\text{mol}^{-1}$ [10] and $V_{Ti} = 10.6\,\text{cm}^3\,\text{mol}^{-1}$ into Eq. (6.18). Therefore, $t_o > t_o^c = 0.51$ holds for the growth of anatase type of TiO_2 film, implying that the

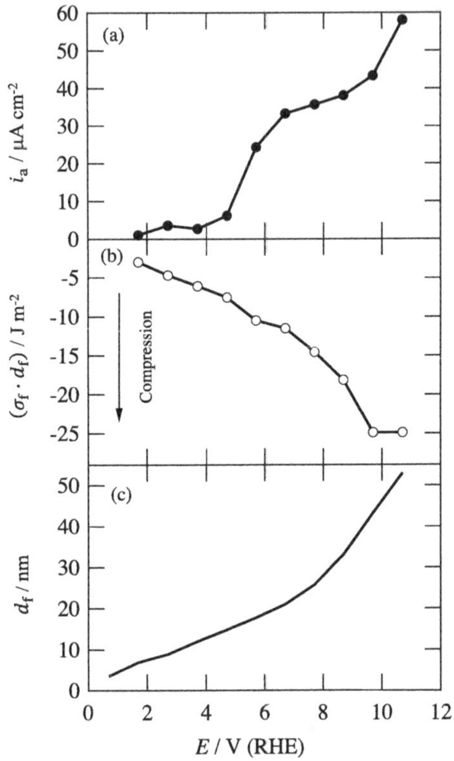

Fig. 6.4 **a** the anodic current density i_a, **b** the stress·thickness product $(\sigma_f \cdot d_f)$, and **c** the thickness of the anodic oxide film d_f at 1 h-anodic oxidation as a function of applied potential E for a sputter-deposited Ti thin film (with a thickness of 250 nm) on a glass plate anodically oxidized by a potential step up to 10.7 V (RHE) in pH 8.4 borate solution [19]. Reproduced from [19] with permission from The Electrochemical Society

observed compressive stress is consistent with that predicted from Nelson–Oriani's criterion [17].

On the contrary, stress variations toward tensile direction have been observed during anodic potential scan of Ti in 0.1 M H_2SO_4 solution [17, 28]. In addition, the compressive stress component of about −190 MPa due to electrostriction was estimated from the stress relaxation during cathodic potential scan in cyclic voltammetry [28]. Moreover, a series of stress measurements during anodic oxidation of Ti thin film at galvanostatic current densities of 0.5–8 mA cm^{-2} up to a cell voltage of 40 V in 1.0 M H_2SO_4 or 0.1 M H_3PO_4 solution [20, 21, 29, 30] revealed the complex behavior of stress variations from compression to tensile during the film growth. Figure 6.5 shows the evolution of the cell voltage V and of the curvature κ of the cantilever electrode during anodic oxidation of a sputter-deposited Ti thin film at a constant current density of 4 mA cm^{-2} in 1.0 M H_2SO_4 solution [20]. As explained

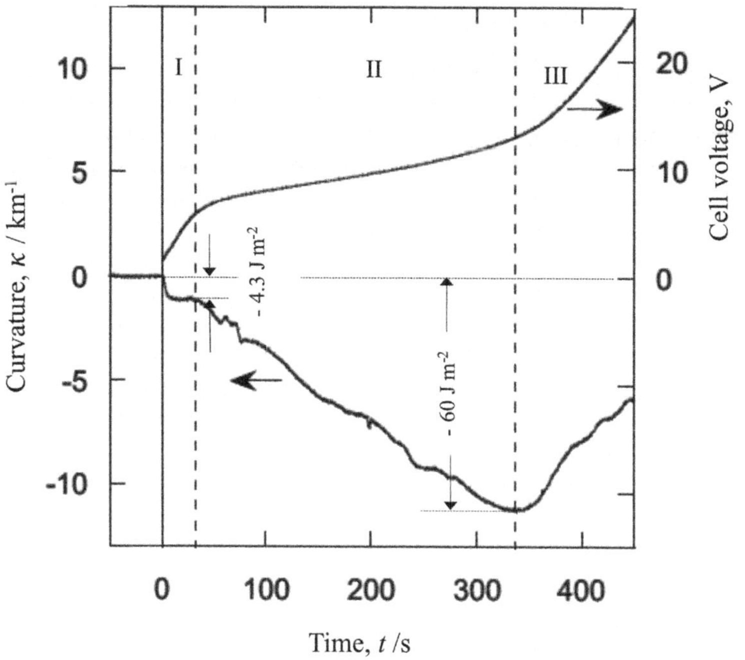

Fig. 6.5 Evolution of the cell voltage V and of the curvature κ of the cantilever electrode during anodic oxidation of a sputter-deposited Ti thin film at a constant current density of 4 mA cm^{-2} in 1.0 M H_2SO_4 solution, exhibiting three distinct stages (I), (II), and (III) [20]. Reproduced from [20] with permission from The Electrochemical Society

in Sect. 2.3 of Chap. 2, stress variations can be measured from the changes in κ. Three distinct stages (I), (II), and (III) can be observed from the changes in V and κ.

At the first stage (I) in Fig. 6.5, κ changes rapidly toward compressive direction at the very beginning of anodic oxidation, accompanying the initial jump of cell voltage, followed by the slight changes of κ toward compressive direction in the cell voltage region (up to 7 V) where the cell voltage rises at a constant rate. The stress·thickness product ($\sigma_f \cdot d_f$) calculated from $\Delta\kappa$ at 5 V is -4.3 J m^{-2}. The separate experiment of the current interruption from 5 V to an open-circuit potential showed the change of $\Delta(\sigma_f \cdot d_f)$ from -4.3 to 0.06 J m^{-2} [20]. The detailed analysis [20] of the curvature data at this stage indicated that the average stress of the film at 5 V is -290 ± 50 MPa which consists of a reversible, field-induced, compressive electrostriction stress of -370 ± 50 MPa and of an irreversible, growth-induced, tensile stress of 79 $\pm$ 20 MPa. The sign of the tensile stress for the irreversible film growth is consistent with the results obtained by an anodic potential scanning or cyclic voltammetry, but it is opposite to the results obtained by a potentiostatic anodic oxidation for 1 h in pH 8.4 borate solution [19] or by scanning repeatedly the cell voltage from 1 to 9 V in 1.0 M H_2SO_4 solution [29].

Crystallization and Oxygen Evolution

At the stage (II) in Fig. 6.5, the slope of the V vs. t curve decreases progressively and then reaches a steady-state value, accompanying the rapid change of κ toward negative direction corresponding to the development of large compressive stress. The decrease in slope of the V vs. t curve may come from the decrease in current efficiency of the film growth due to oxygen evolution associated with the crystallization of the anodic oxide film [27]. The crystallization provides the increase in density of the anodic oxide film that tends to induce the film stress toward tensile direction due to the volume contraction of the film. In the transition from amorphous to anatase type of TiO_2 film, the volume contraction of 5 $cm^3\ mol^{-1}$ is estimated from the difference in density between amorphous ($\rho_{ox} = 3.1\ g\,cm^{-3}$) [31] and anatase type ($\rho_{ox} = 3.84\ g\,cm^{-3}$) [10] of TiO_2 film. Consequently, the observed large compressive stress is contradictory to the tensile stress predicted from the volume contraction due to the crystallization of the anodic oxide film.

Nevertheless, as shown schematically in Fig. 6.6, the formation of nano-crystalline anatase in the inner layer of the anodic oxide film and the development of nanoscale bubbles due to the local generation of oxygen around the nano-crystals have been confirmed by transmission electron microscopic (TEM) observation for the ultra-microtomed section of a sputter-deposited Ti thin film anodized up to a cell voltage of 20 V at a galvanostatic current density of 5 mA cm^{-2} in 0.1 M ammonium pentaborate solution [32]. The transport number of O^{2-} ion $t_o = 0.65$ is determined from the marker position [32]. The outer layer is formed at the film/solution interface by outward migration of Ti^{4+} ($t_{Ti} = 0.35$), while the inner layer is formed at the metal/film interface by inward migration of O^{2-}/OH^- ions ($t_o = 0.65$). The inner

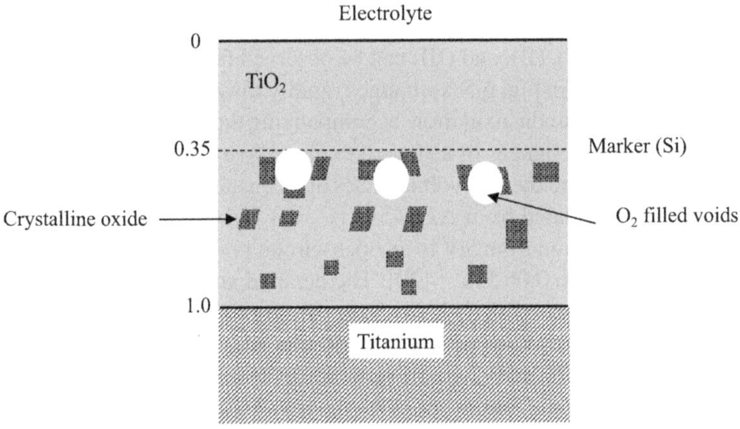

Fig. 6.6 Schematic illustration of the local crystallization and oxygen evolution confirmed by the transmission electron microscopic (TEM) observation for the ultra-microtomed section of a sputter-deposited Ti thin film anodically oxidized up to a cell voltage of 20 V at a galvanostatic current density of 5 mA cm^{-2} in 0.1 M ammonium pentaborate solution [32]. Reprinted from [32], Copyright 2003, with permission from Elsevier

layer contains nano-crystals, and the voids filled with oxygen are formed on nano-crystals within the inner layer near the marker position, indicating that the oxygen evolution occurs on the nano-crystals [32].

The local oxygen evolution requires the transport of the electrolyte to the nano-crystals. Although the mechanism of the electrolyte transport is not clear, it is deduced that the electrolyte is transported to the nano-crystals through flaws or cracks in the outer layer of the film. The local oxygen evolution on the nano-crystals in the inner layer induces the large compressive stress due to local swelling of the inner layer, which predominates over the tensile stress induced by the volume contraction due to local crystallization. At the stage (II), therefore, the increase in compressive stress with time or with cell voltage is ascribed to the local oxygen evolution in the inner layer [20].

At the stage (III), the slope of the V vs. t curve increases and returns to a value close to that at the stage (I), indicating that the oxygen evolution ceases to increase the current efficiency of the film growth. In the response to the increase in current efficiency, the stress varies from compressive to tensile, passing through a maximum compressive value of – 60 J m^{-2} at about 340 s, corresponding to a cell voltage of about 13 V. The kinetics of the local oxygen evolution is limited by the delivery rate of the electrolyte to the nano-crystals, and thus, the oxygen evolution rate reduces as the film thickness, i.e., the length of the transport path increases. If the length of the transport path exceeds a critical value, the local oxygen evolution should cease since the delivery rate of the electrolyte cannot follow the rate of the local oxygen evolution. This may be the plausible reason for the cease of the local oxygen evolution [20]. At a cell voltage larger than 13 V, if the tensile stress induced by the volume contraction due to crystallization of the film prevails over the compressive stress due to the local oxygen evolution, the tensile stress would develop progressively with increasing cell voltage.

Other Factors Influencing Film Stress

Nelson–Oriani's criterion [17] is insufficient to explain the tensile stress of the film at the stage (III) because of $t_o > t_o^c$. In order to satisfy the condition of the tensile stress $t_o < t_o^c$, a crystalline titanium oxide with a density larger than that $\rho_{ox} = 4.23$ g cm^{-3} of the most dense oxide phase, i.e., crystalline rutile has to be assumed, which is obviously unrealistic [24]. Nelson–Oriani's criterion [17] does not account for the defective structure of the oxide film. It is known that the anodic oxide films on Ti have n-type semiconductive properties [26, 33]. Interstitial Ti^{3+} ion or oxygen vacancy are considered to be donors in the anodic oxide film. It has been proposed that the volume contraction of the oxide film due to annihilation of interstitial ion or generation of vacancy induces the tensile stress, while the volume expansion due to generation of interstitial ion or annihilation of vacancy induces the compressive stress [34, 35]. For the anodic oxidation of Al, it has been also pointed out that the changes in sign and magnitude of stress are associated with the annihilation of cation vacancy or the generation of oxygen vacancy at the Al/oxide film interface [36].

Fromhold-Jr theoretically derived that stresses are generated by a momentum exchange due to collision of diffusing defects (interstitial ion or vacancy) with the diffuse media during oxide growth [34, 35]. For example, in the case where interstitial cation diffuses (or migrate) in the oxide film toward the oxide film/solution interface, the lattice constant of the oxide film expands in the direction parallel to the interstitial cation current, leading to the contraction of the lattice constant in the directions perpendicular to the cation current, by which the tensile stress is generated in parallel to the film plane. Unfortunately, there have been no experimental reports of the stress generation associated with the defective structure of the anodic oxide film. As described above, Nelson–Oriani's criterion [17] does not account for the stress factors associated with the annihilation or generation and diffusion (migration) of the defects in the oxide film in addition to the crystallization and oxygen evolution. Nevertheless, it is still worthy at the first diagnosis of the film stress to compare the sign of the measured stress with that predicted from Nelson–Oriani's criterion [17].

6.6 Compressive Stress due to Electrostriction

The electrostriction is one of the important factors influencing the stress of the anodic oxide film. If the anodic oxide film is dielectric, the high electric field across the film during anodic oxidation exerts the stress normal to the film plane (i.e., along the film thickness) due to coulombic attraction between the charges of opposite sign located on the both sides of the dielectric film. If a simple parallel-plate capacitor model is employed, the stress σ_z^{el} along the film thickness [37] is given by

$$\sigma_z^{el} = \left(\frac{\varepsilon_0 \varepsilon_f}{2}\right)\bar{E}^2 = \left(\frac{\varepsilon_0 \varepsilon_f}{2}\right)\left(\frac{\Delta\phi_f}{d_f}\right)^2 \tag{6.26}$$

where ε_0 (= 8.854 × 10^{-12} F m^{-1}) is the vacuum permittivity, ε_f is the relative dielectric constant of the film, $\bar{E}$ is the electric field perpendicular to the film plane, $\Delta\phi_f$ is the potential difference across the film, and d_f is the film thickness. The stress represented by Eq. (6.26) is usually referred as the Maxwell stress [38, 39]. Since the film is mechanically constrained by the metal substrate, σ_z^{el} is converted to the in-plane stress σ_{xy}^{el} parallel to the film plane, i.e., the electrostriction stress in the film. The relationship between σ_z^{el} and σ_{xy}^{el} is given by

$$\sigma_{xy}^{el} = -\left(\frac{\nu_f}{1-\nu_f}\right)\sigma_z^{el}, \tag{6.27}$$

where ν_f is Poisson's ratio of the film. The minus sign in Eq. (6.27) indicates that σ_{xy}^{el} is compressive in the film geometry due to the coulombic attraction along the film thickness.

It has been reported that the electrostriction stresses of the anodic oxide films measured experimentally during anodic oxidation of valve metals are −6.2 M Pa [18] for Al, −37 MPa [28] for Nb, and −190 MPa [28], −240 MPa [29] or −20 MPa

[19] for Ti. The corresponding values of σ_{xy}^{el} calculated by using Eqs. (6.26) and (6.27) are −9.3 MPa (assuming $\nu_f = 0.22$, $\varepsilon_f = 9$, and $\bar{E} = 9.1 \times 10^8\,\mathrm{V\,m^{-1}}$) for Al, –27 MPa (assuming $\nu_f = 0.4$, $\varepsilon_f = 40$, and $\bar{E} = 4.8 \times 10^8\,\mathrm{V\,m^{-1}}$) for Nb and −18 MPa (assuming $\nu_f = 0.25$, $\varepsilon_f = 85$, and $\bar{E} = 3.8 \times 10^8\,\mathrm{V\,m^{-1}}$) for Ti. The experimental values of σ_{xy}^{el} for the anodic oxide films on Al and Nb are close to the calculated values, while the experimental values of σ_{xy}^{el} for the anodic oxide film on Ti except for a small value of −20 MPa [19] are larger by one order of magnitude than the calculated value.

It has been pointed out [29, 40] that Eq. (6.26) does not contain a dielectrostriciton term, and the large difference between experimental and calculated values of σ_{xy}^{el} for the anodic oxide film on Ti results from the neglect of the dielectrostriction term in Eq. (6.26). The dielectrostriction is defined as the variation of the dielectric properties of a material with deformation [41]. The deformation of the anodic oxide film due to applied electric field affects the dielectric constant of the film since the dipoles in the film are aligned along the direction of the applied electric field. In the case of an isotropic material, if the contribution of the dielectrostriction is taken into consideration [42, 43], σ_{xy}^{el} can be expressed by

$$\sigma_{xy}^{el} = -\left(\frac{\nu_f}{1-\nu_f}\right)\left(\frac{\varepsilon_0}{2}\right)[\varepsilon_f - (\alpha_1 + \alpha_2)]\bar{E}^2, \tag{6.28}$$

where α_1 and α_2 are the electrostriction parameters due to changes in film thickness and in its volume, respectively. Furthermore, α_1 and α_2 can be expressed as a function of ε_f:

$$\alpha_1 = -\frac{2}{5}(\varepsilon_f - 1)^2, \tag{6.29}$$

and

$$\alpha_2 = -\frac{1}{3}(\varepsilon_f - 1)(\varepsilon_f + 2) + \frac{2}{15}(\varepsilon_f - 1)^2. \tag{6.30}$$

Consequently, σ_{xy}^{el} is eventually represented by

$$\sigma_{xy}^{el} = -\left(\frac{\nu_f}{1-\nu_f}\right)\frac{\varepsilon_0}{2}\left(0.6\varepsilon_f^2 + 0.8\varepsilon_f - 0.4\right)\bar{E}^2. \tag{6.31}$$

The values of ε_f calculated from Eq. (6.31) by using the experimental values of σ_{xy}^{el} for the anodic oxide film ($\nu_f = 0.25$) on Ti are 54 ± 4 that are statistically equal to those (55 ± 6) obtained independently by impedance measurements [29]. The significantly large electrostriction stress measured in anodic TiO_2 film is explained in terms of the effect of the dielectrostriction. It has been also suggested [29] that the effect of the dielectrostriction for the anodic oxide film with a small relative dielectric constant such as Al_2O_3 ($\varepsilon_f \approx 9$) and Ta_2O_5 ($\varepsilon_f = 27$) films is not significant as compared to that for the anodic oxide film with a large relative dielectric constant such as TiO_2 film ($\varepsilon_f = 60$). A small value of −20 MPa [19] obtained by the cathodic potential scan (1 mV s^{-1}) after the potentiostatic anodic oxidation of Ti for 1 h in pH 8.4 borate solution may result from the compositional alteration of the anodic oxide film such as hydration due to cathodic polarization for a long time.

6.7 Residual Stress of Substrate Metal

The thickness of substrate metal decreases due to growth of the anodic oxide film or anodic dissolution during anodization of the metal. When the substrate metal is subjected to residual stress (compressive or tensile), the stress variations $\Delta(\sigma \cdot d)$ measured during anodic oxidation are influenced by the decrease in thickness of the substrate metal layer as represented by

$$\Delta(\sigma \cdot d) = \sigma_m \Delta d_m + \sigma_f \Delta d_f, \tag{6.32}$$

where σ_m is the residual stress of the substrate metal, $\Delta d_m (< 0)$ the decrease in thickness of the substrate metal layer due to consumption, σ_f the growth stress of the anodic oxide film, and $\Delta d_f (> 0)$ is the increase in film thickness due to film growth. In Eq. (6.32), it is assumed that σ_m and σ_f are uniform in the whole depth of the substrate metal layer and of the anodic oxide film, respectively. Moreover, the contribution of compressive stress component due to electrostriction is removed from Eq. (6.32). Equation (6.32) can be transformed to [20]:

$$\Delta(\sigma \cdot d) = \left(\sigma_m + \sigma_f \frac{\Delta d_f}{\Delta d_m}\right) \Delta d_m = (\sigma_m - \eta_f \alpha_{PB} \sigma_f) \Delta d_m, \tag{6.33}$$

where α_{PB} is the Pilling–Bedworth ratio, and η_f is the formation efficiency of the anodic oxide film. The efficiency of the anodic dissolution of the substrate metal into solution through the anodic oxide film is given by $(1 - \eta_f)$.

It has been reported that the residual stresses of a sputter-deposited Zr film on Si (100) wafer are −80 MPa for a thickness of 235 nm and −40 MPa for a film thickness of 440 nm, respectively [24]. The compressive residual stress for the sputter-deposited film results from an atomic peening due to bombardment of target atoms and ions or inert gas atoms with a kinetic energy of 5–10 eV [44]. In contrast, the residual stresses of evaporated Fe and Cr films with a thickness of 100 nm on MgF_2 at 27 °C are 1.35 GPa and 1.45 GPa (i.e. tensile), respectively [45]. Figure 6.7 shows the stress variations due to the consumption of Fe when a Fe thin film evaporated on a glass plate is subjected to anodic dissolution in pH 8.4 borate solution [46]. Assuming that the anodic dissolution of Fe proceeds with a current efficiency of 100%, i.e., $\eta_f = 0$ in the active dissolution region of Fe, the residual stress σ_{Fe} of the Fe thin film is derived from Eq. (6.33):

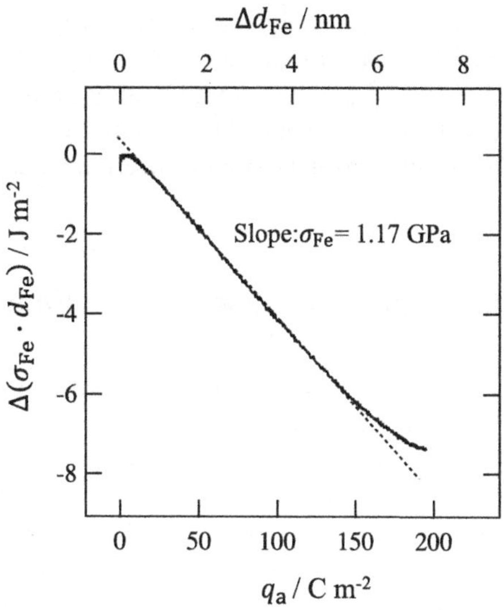

Fig. 6.7 Stress variation $\Delta(\sigma_{Fe} \cdot d_{Fe})$ due to consumption of Fe plotted as a function of electric charge q_a required for the anodic dissolution of Fe or the decrease in film thickness Δd_{Fe} (on the top abscissa) when a Fe thin film evaporated on a glass plate is subjected to anodic dissolution in pH 8.4 borate solution [46]. Reprinted from [46], Copyright 2020, with permission from Springer Nature

$$\sigma_{Fe} = \frac{\Delta(\sigma \cdot d)}{\Delta d_{Fe}}. \tag{6.34}$$

The electric charge q_a (C m^{-2}) required for the anodic dissolution of Fe ($Fe \rightarrow Fe^{2+} + 2e^-$) can be converted to the decrease in film thickness Δd_{Fe}(nm):

$$\Delta d_{Fe} = -\frac{q_a \cdot V_{Fe}}{2F} \times 10^9, \tag{6.35}$$

where $F = 96485$ C mol^{-1} is the Faraday constant and $V_{Fe} = 7.11 \times 10^{-6}$ m^3 mol^{-1} is the molar volume of Fe. The stress changes $\Delta(\sigma_{Fe} \cdot d_{Fe})$ measured by a cantilever bending method are plotted as a function of q_a or Δd_{Fe} (on the top abscissa) in Fig. 6.7 [46]. The value of $\sigma_{Fe} = 1.17$ GPa is obtained from the slope of the linear relation between $\Delta(\sigma_{Fe} \cdot d_{Fe})$ and q_a or Δd_{Fe} represented by the dotted line in Fig. 6.7, which is close to that ($\sigma_{Fe} = 1.35$ GPa) of the evaporated Fe thin film on MgF_2 [45]. In the case of $\eta_f = 1$, Eq. (6.33) is simplified as follows [20]:

$$\Delta(\sigma \cdot d) = (\sigma_m - \alpha_{PB}\sigma_f)\Delta d_m = \left(\sigma_f - \frac{\sigma_m}{\alpha_{PB}}\right)\Delta d_f. \tag{6.36}$$

If σ_m is known, σ_f can be determined from the relationship between $\Delta(\sigma \cdot d)$ and Δd_f by using Eq. (6.36). In the case where the substrate metal is subjected to the residual stress, the relaxation of the residual stress caused by the growth of the anodic oxide film in addition to the anodic dissolution of the substrate metal has to be taken into account to evaluate exactly the film stress [20, 46]. Particularly, if the residual stress of the substrate metal is significantly large, the neglect of the relaxation of the residual stress leads to the major mistakes for the determination of magnitude or sign of the film stress.

6.8 Cathodic Polarization of Anodic Oxide Film

It has been reported [47] that a stress variation toward compressive direction is brought by a cathodic polarization after the formation of anodic oxide film on a sputter-deposited Ti thin film under a potentiostatic anodic oxidation in pH 8.4 borate solution. Figure 6.8 shows (a) the cyclic potential steps, starting from 5.7 to 0.5 V (RHE), then repeating between - 0.5 V and 0.5 V (RHE), and finally returning from −0.5 to 5.7 V (RHE), (b) the corresponding changes in current density, *i*, and (c) the corresponding stress variations ($\sigma_f \cdot d_f$) for the anodic oxide film formed on the sputter-deposited Ti thin film for 1 h at 5.7 V (RHE) in pH 8.4 borate solution [47]. After each potential step, the potential is kept for 1 h to observe the steady-state behavior. The stress varies toward compressive direction during the formation of anodic oxide film at 5.7 V (RHE), which is consistent with previous results obtained under the same anodic polarization conditions for the sputter-deposited Ti thin film [19]. The ellipsometrical thickness of the anodic oxide film formed on Ti for 1 h at 5.7 V (RHE) is about 18 nm, irrespective of kinds of electrolytes used in experiments [5].

At the first potential step from 5.7 to −0.5 V (RHE), ($\sigma_f \cdot d_f$) passes through an instantaneous change toward tensile direction and then changes toward compressive direction. The instantaneous change may result from the relaxation of the electrostriction stress component. The final changes of ($\sigma_f \cdot d_f$) toward compressive direction result from the compositional changes of the anodic oxide film on Ti due to cathodic polarization at −0.5 V (RHE). The compositional changes of the anodic oxide film on Ti during cathodic polarization in pH 6.9 phosphate solution have been investigated by ellipsometry [48]. It has been proposed that the following compositional changes of the anodic oxide film, accompanying hydrogen absorption, take place during cathodic polarization [48]:

$$\mathrm{TiO_2} + x\mathrm{H^+_{aq}} + x\mathrm{e^-} = \mathrm{TiO_{2-x}(OH)_x}, \tag{6.37}$$

where x varies from 0 to 1 with changing cathodic potential from −0.25 to −0.9 V (RHE). Simultaneously, the complex refractive index of the oxide film, $N_f = n - ki$ (n: refractive index, k: extinction index) changes from $N_f = 2.1 - 0.03i$ to $1.85 - 0.35i$ [48]. In addition, the valence change of titanium ions has to be taken into consideration to maintain the charge neutrality during hydrogen ingress. Therefore, Eq. (6.37) can be rewritten by introducing the valence change of titanium ions [47]:

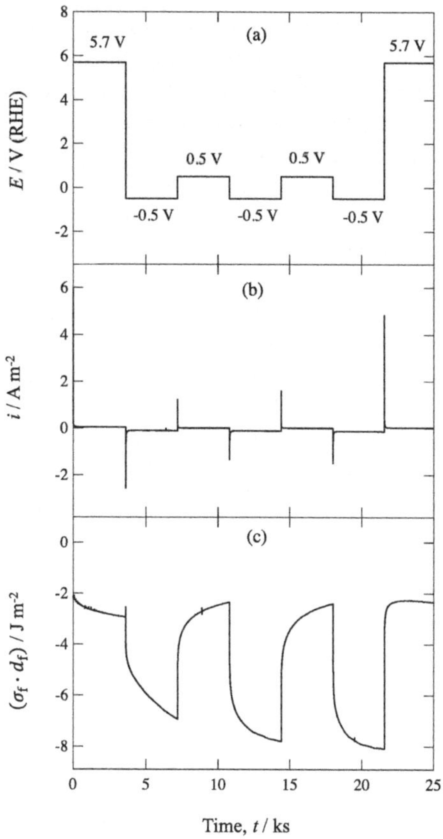

Fig. 6.8 **a** The cyclic potential steps, starting from 5.7 to −0.5 V (RHE), then repeating between −0.5 and 0.5 V (RHE), and finally returning from −0.5 to 5.7 V (RHE), **b** the corresponding changes in current density i, and **c** the corresponding stress changes ($\sigma_f \cdot d_f$) for the anodic oxide film formed on the sputter-deposited Ti thin film for 1 h at 5.7 V (RHE) in pH 8.4 borate solution [47]. After each potential step, the potential is kept constant for 1 h to observe the steady-state behavior. Modified from [47], Copyright 2003, with permission from Elsevier

$$TiO_2 + xH^+_{aq} + xe^- = \left[xTi^{3+}(1-x)Ti^{4+}\right]O_{2-x}(OH)_x. \qquad (6.38)$$

According to Michaelis et al. [49], the molar volume (29.42 cm^3 mol^{-1}) of TiOOH ($x = 1$) is larger than that (24.97 cm^3 mol^{-1}) of TiO_2 ($x = 0$), which reflects on the difference in refractive index between TiOOH ($n = 1.85$) and TiO_2 film ($n = 2.1$). Consequently, it is obvious that the final changes of ($\sigma_f \cdot d_f$) toward compressive direction at the first potential step from 5.7 to −0.5 V (RHE) are caused by the volume expansion due to the compositional changes of the oxide film.

The subsequent potential step from −0.5 to 0.5 V (RHE) causes the anodic current transient and the changes of ($\sigma_f \cdot d_f$) toward tensile direction. The further potential step from 0.5 to −0.5 V (RHE) reverses the current transient and the changes of ($\sigma_f \cdot d_f$).

The reversible current transient and changes of $(\sigma_f \cdot d_f)$ in response to the cyclic potential steps between −0.5 and 0.5 V (RHE) are originated from the reversible reaction of Eq. (6.37) or Eq. (6.38) accompanying the volume expansion and contraction of the oxide film due to the hydrogen absorption and desorption. The steady-state level of $(\sigma_f \cdot d_f)$ after the potential step from −0.5 to 0.5 V (RHE) does not deviate significantly from that after the final potential step from −0.5 to 5.7 V (RHE), indicating that the hydrogen absorbed in the oxide film at −0.5 V (RHE) is completely desorbed from the film at 0.5 V (RHE).

The difference $\Delta\sigma_f = 0.32$ GPa in film stress between TiOOH and TiO_2 was calculated by assuming the film thickness of 18 nm and the molar volume (20.8 cm^3 mol^{-1}) [10] for the anatase type of TiO_2 from the linear relation between $\Delta(\sigma_f \cdot d_f)$ and anodic charge ΔQ_a (for hydrogen desorption) obtained during potential steps from various cathodic potentials (−0.7 ~ −0.2 V) to 0.5 V (RHE) after the formation of the TiO_2 film for 1 h at 5.7 V (RHE) [47]. It has been also reported [50] that downward and upward potential steps after the formation of the passive film on Ni induce the alternative changes of $(\sigma_f \cdot d_f)$ toward compressive and tensile direction, which was ascribed to the volume expansion and contraction of the film due to hydrogen absorption and desorption, accompanying the valence changes between Ni^{2+} and Ni^{3+} in NiO_{1+x} film with cation vacancies. Moreover, the stress variations toward compressive direction have been observed during cathodic reduction of the passive film (γ-Fe_2O_3 or γ-FeOOH) on Fe in pH 8.4 borate solution [46], which was ascribed to the volume expansion due to $Fe(OH)_2$ formed as an intermediate step of the cathodic reduction.

6.9 Plastic Flow of Porous Anodic Oxide Film

It is known that porous alumina films are formed by anodic oxidation of Al in acid solutions [51]. The porous films consist of a thin barrier layer next to the metal substrate and of an outer porous layer with pores of approximately cylindrical section as schematically illustrated in Fig. 6.9 [52]. Figure 6.10 shows the scanning electron microscopic (SEM) images of (a) the substrate surface after removal of the porous alumina film and of (b) the cross section of the porous alumina film formed on Al at a cell voltage of 240 V for 1 h in 2 M citric acid [53]. The SEM images indicate that the porous alumina film has self-organized hexagonal pore arrays. The growth mechanism of the porous alumina film in phosphoric acid [52, 54] has been investigated by monitoring the motion of tungsten tracer layer (3–5 nm thick) in a sputter-deposited Al thin film on substrate Al. Tungsten is an ideal tracer since the W^{6+} ion moves outward in the anodic alumina film much slower than Al^{3+} ion [55].

The upper part of Fig. 6.11 shows the transmission electron microscopic (TEM) images for the cross sections of the sputter-deposited aluminum with an incorporated tungsten tracer layer, followed by anodic oxidation for (a) 180 s, (b) 240 s, and (d) 350 s at 5 mA cm^{-2} in 0.4 M phosphoric acid solution [52, 54]. The lower part of Fig. 6.11 shows schematically the relative distribution of the tungsten tracer layer in alumina films corresponding to the TEM cross sections [54]. The TEM cross section

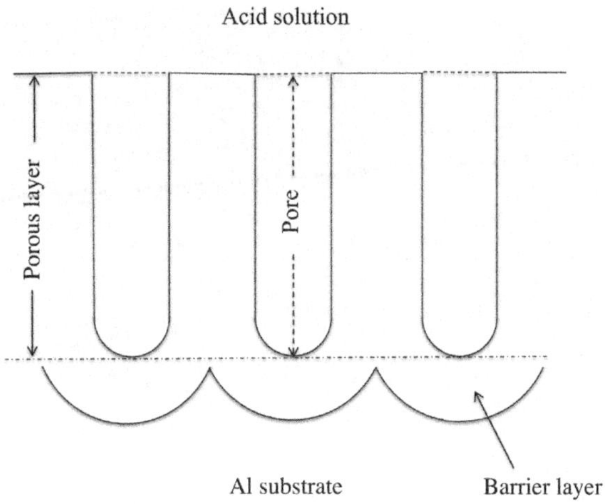

Fig. 6.9 Schematic illustration for the structure of the porous alumina film formed by anodic oxidation of Al in acid solutions [52]. The porous films consist of a thin barrier layer next to the metal substrate and of an outer porous layer with pores of approximately cylindrical section. Reproduced from [52] with permission from The Electrochemical Society

of the Al specimen anodically oxidized for (a) 180 s, corresponding to the start of the tracer oxidation, indicates that the tungsten tracer layer (dark band due to atomic number contrast) just enters the film beneath pore base, while it remains still in the substrate near the cell boundary region. The tracer in the film is located up to about 30% of the thickness of the barrier layer (about 115 nm). The thickness (430 ± 35 nm) of the porous layer region is larger by a factor of about 1.4 than the thickness (311 nm) of the Al substrate consumed by the anodic oxidation.

At the anodic oxidation for (b) 240 s, the tracer band is situated in the mid-thickness within the barrier layer region [52, 54]. It is noted that the tracer band is severely distorted in traversing a cell and the tracer at the cell wall region is about 70–80 nm above the tracer beneath the pore. Furthermore, the tracer near the cell boundaries falls sharply toward the middle of the barrier layer beneath the pore where the band is faint due to the decrease in concentration of tungsten [52, 54]. At the anodic oxidation for (d) 350 s, the tracer band at the cell wall region is located at a depth of 38–50% of the thickness (about 770–800 nm) in the porous layer region and is at least more than about 300 nm above the tracer beneath the pore, although it is not clearly observed because the band is much fainter. The schematic distribution of (c) 300 s in the lower part of Fig. 6.11 assumes a similar displacement of the tungsten tracer layer during film growth for a further 60 s after 240 s, implying that no more than a few percent of the original tungsten should reach eventually the pore base, which is consistent with the negligible losses of tungsten species during anodic oxidation, indicated by Rutherford backscattering spectroscopy (RBS) [54].

According to the conventional field-assisted dissolution model of the porous film growth [56], the tracer is incorporated initially into the film at locations beneath

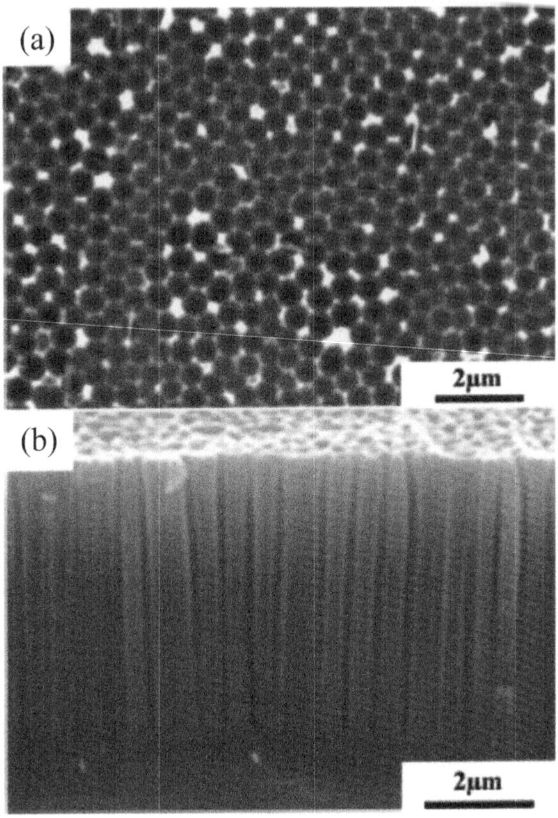

Fig. 6.10 Scanning electron microscopic (SEM) images of **a** the substrate surface after removal of the porous alumina film and of **b** the cross section of the porous alumina film formed on Al at a cell voltage of 240 V for 1 h in 2 M citric acid [53]. The SEM images indicate that the porous alumina film has self-organized hexagonal pore arrays. Reproduced from [53] with permission from The Electrochemical Society

the pores, where the scalloped metal/film interface first intersects the tracer band, and the final incorporation takes place at locations of the cell boundaries. Since the tungsten migrates slowly outward within the alumina film, the tungsten first incorporated should lie ahead of that finally incorporated into the film. However, the real distribution of tungsten species in the film is inverted with respect to expectations of the conventional model. In contrast to the conventional model, Skeldon et al. [52] proposed that the thickness of the barrier layer is kept constant by a plastic flow of oxide from the barrier layer toward the cell wall, which is driven by the compressive stress due to the electrostriction at the pore base and/or by the volume expansion due to oxidation. The significant increase in thickness of the porous layer region relative to that of the metal consumed can be explained only in terms of such plastic flow.

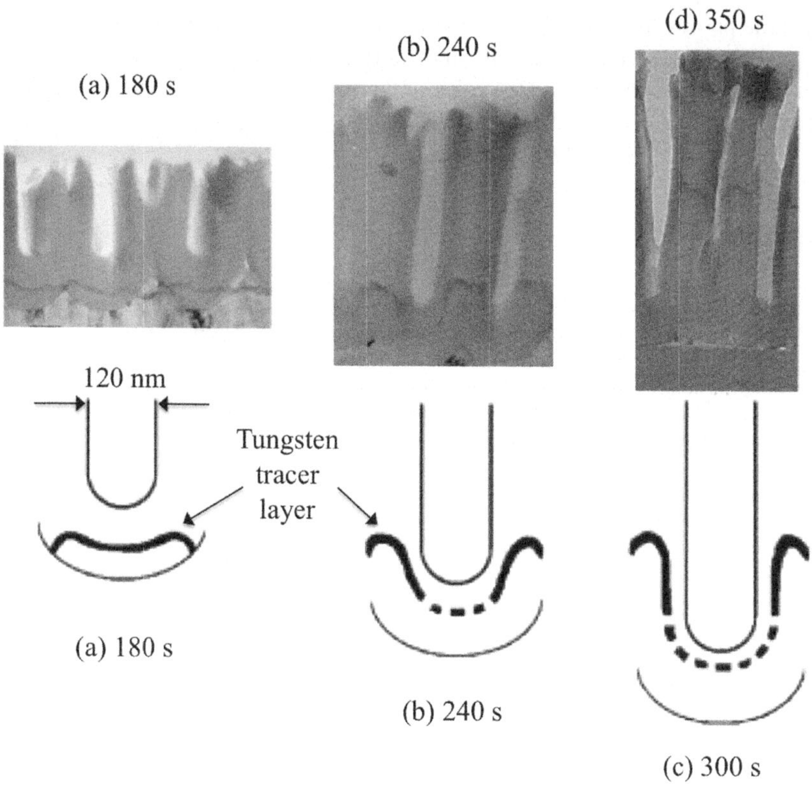

Fig. 6.11 Upper part: transmission electron microscopic (TEM) images of the cross sections of the sputter-deposited aluminum with an incorporated tungsten tracer layer, followed by anodic oxidation for **a** 180 s, **b** 240 s, and **d** 350 s at 5 mA cm^{-2} in 0.4 M phosphoric acid solution [52, 54], and lower part: schematic diagrams showing the relative distribution of the tungsten trace layer in alumina films at each interval of 60 s during anodic oxidation at 5 mA cm^{-2} in 0.4 M phosphoric acid solution: **a** 180 s, **b** 240 s, and **c** 300 s [54]. The distribution of **c** 300 s in the lower figure assumes a similar displacement of the tungsten trace layer during film growth for a further 60 s after 240 s. Reproduced from [52] with permission from The Electrochemical Society, and reprinted from [54], Copyright 2006, with permission from Elsevier

The plasticity in alumina films during anodic oxidation is compatible with the observations of film plasticity [13, 57, 58]. The compressive stress of about –150 MPa responsible for the plastic flow of the barrier layer is estimated from the growth of oxygen gas bubbles observed in the barrier alumina film during anodic oxidation of Al-0.47 atomic % Au alloy [57]. The stress level of an order of –100 MPa is also estimated for the electrostriction, which is sufficient to deform oxides [59]. The plastic flow is limited to the region of ionic transport in the barrier layer and is directly associated with the migration process. In addition, it is emphasized [52] that electrolyte anions such as phosphate or sulfate ions incorporated in the barrier layer influence the electric field, the film plasticity, and the resultant deformation behavior since the porous films formed in borax and chromic acid solutions contain no electrolyte species and have a less regular morphology. At present, however, further work is needed to understand the roles of the incorporated electrolyte anions in the formation of self-organized pore arrays.

References

1. Cabrera N, Mott NF (1948–1949) Rep Progr Phys 12:163–184
2. Young L (1961) Anodic oxide films. Academic Press, New York, pp 13–37
3. Lohrengel MM (1993) Mater Sci and Eng R 11:243–294
4. Guntherschulze A, Betz H (1937) Die elektrolyt kondensatoren. Verlag M Klain, Berlin
5. Ohtsuka T, Masuda M, Sato N (1985) J Electrochem Soc 132:787–792
6. Khalil N, Leach JSL (1986) Electrochim Acta 31:1279–1285
7. Vijh AK (1973) Electrochemistry of metals and semiconductor: the application of solid state science to electrochemical phenomena. Marcell Dekker Inc, New York, pp 129–165
8. Pilling NB, Bedworth RE (1923) J Inst Met 29:529–591
9. Dictionary of science and chemistry (Rikagaku-jiten in Japanese) (1987) Iwanami Shoten Publishers, Tokyo
10. Samsonov GV (1973) The oxide handbook. IFI/Plenum, New York
11. Davies JA, Domeij B, Pringle JPS, Brown F (1965) J Electrochem Soc 112:675–680
12. Amsel G, Samuel D (1962) J Phys Chem Solids 23:1707–1718
13. Pringle JPS (1980) Electrochim Acta 25:1403–1428
14. Cherki C, Siejka J (1973) J Electrochem Soc 120:784–791
15. Siejka J, Ortega C (1977) J Electrochem Soc 124:883–891
16. Brown F, Mackintosh WD (1973) J Electrochem Soc 120:1096–1102
17. Nelson JC, Oriani RA (1993) Corros Sci 34:307–326
18. Capraz ÖÖ, Shrotriya P, Hebert KR (2014) J Electrochem Soc 161:D256–D262
19. Ueno K, Pyun S-I, Seo M (2000) J Electrochem Soc 147:4519–4523
20. Vanhumbeek J-F, Proost J (2008) J Electrochem Soc 155:C506–C514
21. Van Overmeere Q, Vanhumbeek J-F, Proost J (2010) J Electrochem Soc 157:C166–C173
22. Vermilyea DA (1963) J Electrochem Soc 110:345–346
23. Leach JSL, Pearson BR (1988) Corros Sci 28:43–56
24. Van Overmeere Q, Proost J (2010) Electrochim Acta 55:4653–4660
25. Bradhurst DH, Leach JSL (1966) J Electrochem Soc 113:1245–1249
26. Ohtsuka T, Otsuki T (1988) Corros Sci 40:951–958
27. Ohtsuka T, Guo J, Sato N (1986) J Electrochem Soc 133:2473–2476
28. Sahu SN, Scarminio J, Decker F (1990) J Electrochem Soc 137:1150–1154
29. Vanhumbeeck J-F, Proost J (2008) Electrochim Acta 53:6165–6172
30. Proost J, Vanhumbeeck J-F, Van Overmeere Q (2009) Electrochim Acta 55:350–357
31. Dyer CK, Leach JSL (1978) J Electrochem Soc 125:1032–1038
32. Habazaki H, Uozumi M, Konno H, Shimizu K, Skeldon P, Thompson GE (2003) Corros Sci 45:2063–2073
33. Blackwood DJ, Peter LM (1989) Electrochim Acta 34:1505–1511
34. Fromhold-Jr AT (1972) Surf Sci 22:396–410
35. Fromhold-Jr AT (1976) Theory of metal oxidation. vol 1. Fundamentals. North-Holland Pub Co, Amsterdam, pp 228–230
36. Moon S-M, Pyun S-I (1998) Electrochim Acta 43:3117–3126
37. Butler MA, Ginley DS (1988) J Electrochem Soc 135:45–51
38. Landau LD, Lifshitz EM (1960) Electrodynamics of continuous media: course of theoretical physics. vol 8. Pergamon, Oxford, pp 31

39. McMeeking RM, Landis CM (2005) J Appl Mech 72:581–590
40. Vanhumbeeck J-F, Proost J (2007) Colloid Surf B: Biointerf 56:163–169
41. Peng Y, Shkel YM, Kim G-H (2005) J Rheol 49:297–311
42. Shkel YM, Klingenberg DJ (1998) J Appl Phys 83:7834–7843
43. Lee HY, Peng Y, Shkel YM (2005) J Appl Phys 98:074104–1–9
44. D'Heurle FM, Harper JME (1989) Thin Solid Films 171:81–92
45. Thurner G, Aberman R (1990) Thin Solid Films 192:277–285
46. Seo M, Ueno K (2020) J Solid State Electrochem 24:929–940
47. Kim J-D, Pyun S-I, Seo M (2003) Electrochim Acta 48:1123–1130
48. Ohtsuka T, Masuda M, Sato N (1987) J Electrochem Soc 134:2406–2410
49. Michaelis A, Delplancke JL, Schultze JW (1995) Mater Sci Forum 185–188:471–480. (Ellipsometric determination of density of TiO_2 passive films on Ti single crystals: combination of ellipsometry and coulometry. In: Heusler KE (ed) Proc 7 th international symposium on passivity. Passivation of metals and semiconductors. Trans Tech Publications Ltd, Andermannsdorf, Switzerland)
50. Kim J-D, Seo M (2003) J Electrochem Soc 150:B193–B198
51. Thompson GE, Wood GC (1983) Anodic film on aluminium. In: Scully JC (ed) Corrosion: aqueous processes and passive films. Academic Press, New York, pp 205–329
52. Skeldon P, Thompson GE, Garcia-Vergara SJ (2006) Eletrochem Solid State Lett 9:B47–B51
53. Ono S, Saito M, Ishiguro M, Asoh H (2004) J Electrochem Soc 151:B473–B478
54. Garcia-Vergara SJ, Skeldon P, Thompson GE, Habazaki H (2006) Electrochim Acta 52:681–687
55. Habazaki H, Shimizu K, Skeldon P, Thompson GE, Wood GC (1996) J Electrochem Soc 143:2465–2470
56. Hoar TP, Mott NF (1959) J Phys Chem Solids 9:97–99
57. Zhou Z, Habazaki H, Shimizu K, Skeldon P, Thompson GE, Wood GC (1999) Proc R Soc London Ser A 455:385–399
58. Iglesias-Rubianes L, Skeldon P, Thompson GE, Habazaki H, Shimizu K (2002) J Electrochem Soc 149:B23–B26
59. Sato N (1971) Electrochim Acta 16:1683–1692
60. DiQuarto F, Doblhofer K, Gerischer H (1987) Electrochim Acta 23:195–201
61. Stringer J (1970) Corros Sci 10:513–543
62. Archibald LC (1977) Electrochim Acta 22:657–659
63. Archibald LC, Leach JSL (1977) Electrochim Acta 22:15–20

Chapter 7
Nano-Mechanical Properties of Solid Surfaces Obtained by Nano-Indentation

Abstract Nano-indentation is a powerful method for investigating mechanical properties such as hardness, elastic modulus, and yielding strength in a nanometer's range (i.e. nano-mechanical properties) of solid surfaces. The nano-mechanical properties can be determined from the load-depth curves measured by nano-indentation. The principle, instrumentation, and features of nano-indentation are described as fundamentals. The nano-mechanical properties of bare metal surface, metal oxide surface, and thin oxide film on metal are explained as typical examples, and the issues for determination of hardness from the measured load-depth curve in addition to the indentation size effect on hardness are discussed. Electrochemical nano-indentation can be achieved during a potentiostatic polarization of a solid electrode in electrolyte solution. The hardness values of passive single crystal iron surfaces obtained by electrochemical nano-indentation are increased by chromate treatment of the iron surfaces, which is explained in terms of the promotion of corrosion resistivity due to chromate treatment.

Keywords Nano-mechanical properties · Nano-indentation · Load-depth curve · Hardness · Elastic modulus

7.1 Introduction

Nano-indentation is a powerful method for investigating the mechanical properties such as hardness, elastic modulus, and yielding strength in a nanometer's range (i.e. nano-mechanical properties) of solid surfaces. The nano-mechanical properties of the sample surfaces without preexisting defects can be evaluated from the load-depth curves measured by nano-indentation under a maximum load of 10^{-5}–10^{-4} N because of a small indentation volume (10^{-6}–10^{-3} μm^3). It is known that the hardness value obtained by nano-indentation differs from that by micro-indentation even if the same sample surface is tested. The hardness values obtained by nano- or micro-indentation depend on the indentation depth, which is named "indentation size effect (ISE)".

Although the nano-indentation test has been mostly performed for solid surfaces in air or in vacuum, its test can be applied to a solid electrode in electrolyte solution by using a miniature electrochemical cell. The latter is named "electrochemical

M. Seo, *Electro-Chemo-Mechanical Properties of Solid Electrode Surfaces*,
https://doi.org/10.1007/978-981-15-7277-7_7

nano-indentation" to distinguish with the former usual test. Electrochemical nano-indentation can be achieved during a potentiostatic polarization of a solid electrode.

The present chapter deals with the nano-mechanical properties of bare metal surface, bulk metal oxide surface, and thin oxide film on metal obtained by nano-indentation. We discuss the issues for determination of hardness from the measured load-depth curve in addition to the indentation size effect on hardness. Furthermore, we discuss the nano-mechanical properties of passive single crystal iron surfaces in solution, obtained by electrochemical nano-indentation.

7.2 Fundamentals of Nano-Indentation

The nano-indentation apparatus combined with atomic force microscope (AFM) is commercially available. The original AFM head (cantilever and detector) is replaced by a three-plate capacitive force/displacement transducer with a shaft (tungsten rod) on which an indenter tip is mounted. The transducer can generate the electrostatic force used for the indentation and simultaneously measure the relationship between load (force) and displacement of the indenter [1]. Since the transducer with the indenter tip is mounted on the AFM, the same tip can serve as both indenter and scanning probe, which allows the characterization of the sample surface prior to the indentation and the immediate imaging after the load removal.

Figure 7.1 shows the geometries of diamond tips mostly employed for nano-indentation test. A Berkovich tip (see Fig. 7.1a) has a total included angle (A + B) of 142.3° with a centerline-to-face angle (B) of 65.35°. The curvature radius of the Berkovich tip is 100–200 nm. A cube-corner tip (see Fig. 7.1b) has a total included angle (A + B) of 90° with a centerline-to-face angle (B) of 35.3°. The curvature radius (40–60 nm) of the cube-corner tip is much smaller than that of the Berkovich tip. The cube-corner tip because of sharp angle and high aspect ratio is specified to ultra-thin films. A cono-spherical (conical) tip with a spherical end (see Fig. 7.1c) has an included(cone) angle (C) of 140.6° or 90°, and the curvature radius (1–3 μm) of the tip end is larger than that of the Berkovich tip. The indentation tip used for electrochemical nano-indentation test on a solid electrode in liquid is mounted on a long shaft which extends the tip away from the transducer to avoid the contact of the transducer with liquid.

For nano-indentation test, the load is applied at a constant loading velocity vertically to a flat sample surface by an indenter tip, and simultaneously the displacement of the sample surface in depth direction (depth-displacement) under the indenter is measured as a function of load. After attaining a maximum load, the load is released at a constant unloading velocity and the recovery in depth-displacement is measured during unloading. After the indentation, the surface profile of the indent can be observed from the atomic force microscopic (AFM) image measured by using the same indenter tip. Figure 7.2 represents schematically a load-depth (displacement) curve for nano-indentation test. During loading, the sample surface region of very small volume under the indenter is elastically and then plastically deformed as the

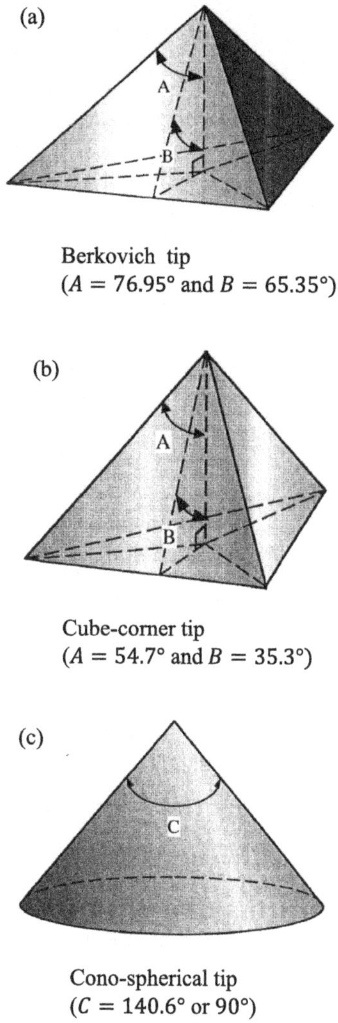

Fig. 7.1 Geometries of diamond tips mostly employed for nano-indentation test: **a** Berkovich tip, **b** cube-corner tip, and **c** cono-spherical (conical) tip

contact area increases with increasing depth-displacement. The part of the unloading curve is dominated by elastic displacement. The mechanical properties such as hardness, elastic modulus, and yield strength of the solid surface region in confined small volume can be obtained from the measured load-depth (displacement) curve based on the method developed by Oliver and Pharr (named "Oliver–Pharr method") [1, 2].

The hardness H is defined as the maximum load $L_{\max}$ divided by the projected contact area A of the indenter at $L_{\max}$ [1–3]:

$$H = \frac{L_{\max}}{A}. \tag{7.1}$$

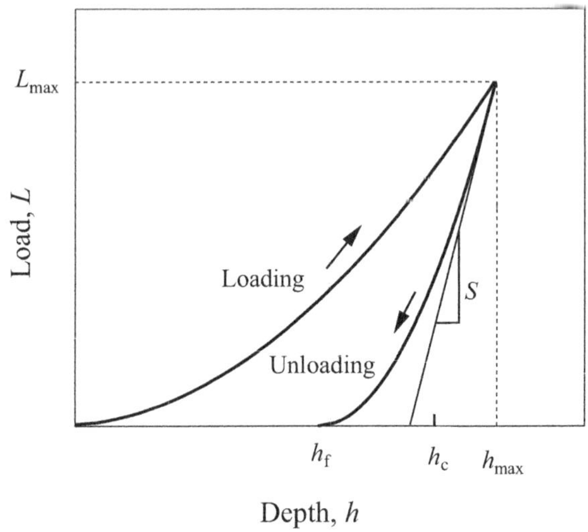

Fig. 7.2 Schematic illustration of a load-depth curve for nano-indentation test [1, 2]. The symbols shown are L_{max}: the maximum indentation load; h_{max}: the displacement of the indenter at L_{max}; h_c : the contact depth of the indenter with the sample surface at L_{max}; S: the stiffness of the sample surface at initial unloading; and h_f: the final depth of the contact surface after unloading. Reprinted from [1], Copyright: Materials Research Society 1992, by permission from Cambridge University Press

If the indenter has an ideal geometry, the projected contact area A can be obtained from the contact depth h_c at L_{max}:

$$A = c_o h_c^2, \tag{7.2}$$

where $c_o = 24.5$ for a Berkovich tip and $c_o = 2.98$ for a cube-corner tip. As seen from Fig. 7.3 which represents schematically a cross section of the surface profile at nano-indentation, the contact depth h_c differs from the indentation depth h_{max} at L_{max} since the contact perimeter of the indenter is deformed by h_s. Consequently, h_c is given by

$$h_c = h_{max} - h_s. \tag{7.3}$$

Furthermore, the following relationship [1] is derived on the basis of Sneddon's analysis [4] of the shape of the surface outside the contact area:

$$h_c = h_{max} - \varepsilon \frac{L_{max}}{S}, \tag{7.4}$$

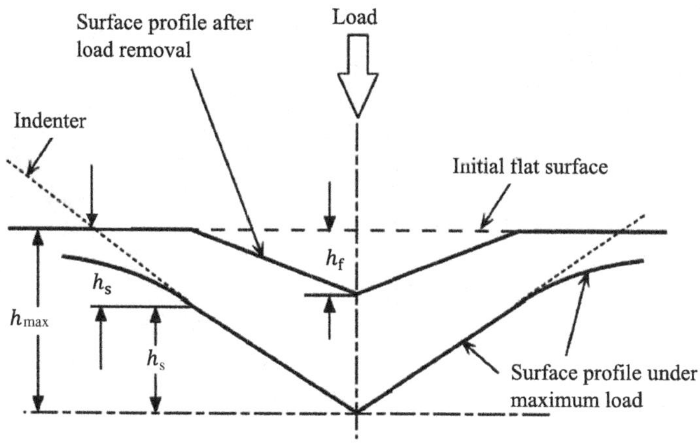

Fig. 7.3 Schematic illustration of a cross section of surface profile at nano-indentation [1, 2]. The symbols shown are same as those in Fig. 7.2. Reprinted from [1], Copyright: Materials Research Society 1992, by permission from Cambridge University Press

where ε is the geometrical constant depending on the shape of the indenter and S is the stiffness (the reciprocal of the compliance) which is equal to the slope of the unloading curve at L_{max} in the measured load-depth curve. The values of $\varepsilon = 0.75$ and 0.72 are given for the Berkovich or cube-corner indenter and for the conical indenter, respectively.

For exact determination of the hardness, the shape calibration of the indenter tip is dispensable since the geometry of the indenter tip is usually not ideal. and besides, the tip becomes blunt due to wear during repeated experiments for a long term. The tip shape calibration is based on determination of the area function $A(h_c)$, which relates the projected contact area A to the contact depth h_c. Fused silica with known mechanical properties (Young's modulus and Poisson's ratio) is used as a standard sample for calibration purpose. Figure 7.4 shows the load-depth curves of the fused silica measured at various maximum loads by using a Berkovich indenter. Assuming that Young's modulus of the fused silica is constant, independent of the indentation depth, the following relationship holds between the projected contact area A and stiffness S:

$$A = \left(\frac{\pi}{4}\right)\left(\frac{S}{E_r}\right)^2, \tag{7.5}$$

where E_r is the reduced modulus. The reduced modulus E_r is given by

$$\frac{1}{E_r} = \frac{\left(1 - \nu_s^2\right)}{E_s} + \frac{\left(1 - \nu_i^2\right)}{E_i}, \tag{7.6}$$

Fig. 7.4 Load-depth curves of the fused silica measured at various maximum loads by using a Berkovich indenter

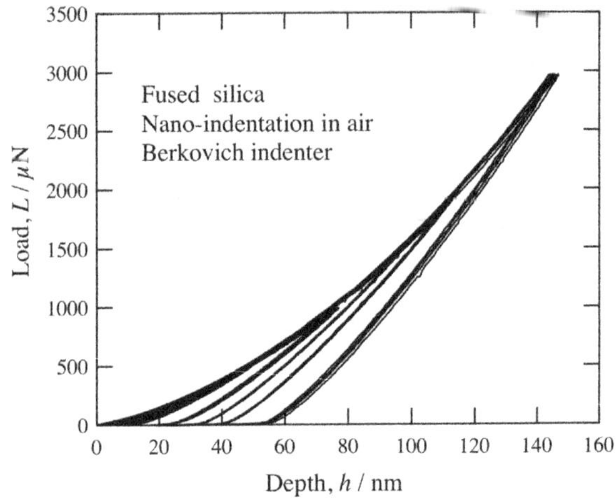

where E_s and ν_s are Young's modulus and Poisson's ratio for the fused silica, and E_i and ν_i are the same parameters for the indenter. The value of $E_r = 69.6$ GPa is obtained from Eq. (7.6) in the case where a pyramidal diamond tip ($E_i = 1140$ GPa and $\nu_i = 0.07$) is used as an indenter for the fused silica ($E_s = 72$ GPa and $\nu_s = 0.17$). The area function $A(h_c)$ is eventually obtained by plotting the projected contact area A calculated from Eq. (7.5) versus the contact depth h_c calculated from Eq. (7.4).

Figure 7.5 shows the projected contact area A obtained as a function of h_c for the fused silica from the load-depth curves in Fig. 7.4. In Fig. 7.5, $\varepsilon = 0.75$ is employed

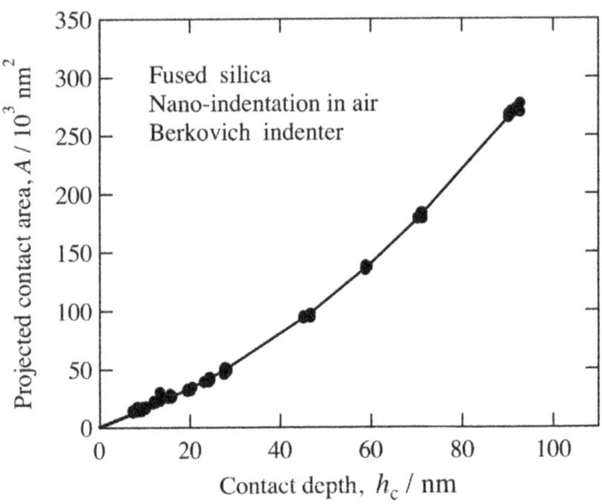

Fig. 7.5 Projected contact area A obtained as a function of contact depth h_c for the fused silica from the load-depth curves in Fig. 7.4. In Fig. 7.5, $\varepsilon = 0.75$ is employed for the calculation of h_c from Eq. (7.4)

for the calculation of h_c from Eq. (7.4). The area function can be fitted to a fifth-order polynomial form:

$$A = c_0 h_c^2 + c_1 h_c + c_2 h_c^{\frac{1}{2}} + c_3 h_c^{\frac{1}{4}} + c_4 h_c^{\frac{1}{8}} + c_5 h_c^{\frac{1}{16}}, \tag{7.7}$$

where $c_i (i = 0, 1, 2, 3, 4, 5)$ are constants, independent of h_c, and the fixed values of $c_0 = 24.5$ and 2.98 are used for the Berkovich and cube-corner indenters, respectively. The values of c_i in Eq. (7.7) except for c_0 are determined by curve fitting with the A vs h_c curves shown in Fig. 7.5. Once the values of c_i are determined, the projected contact area A at any h_c is obtained from Eq. (7.7), and thus, the hardness H can be determined as a function of h_c from Eq. (7.1). Equation (7.5) is rewritten by

$$S^{-1} = \frac{\sqrt{\pi}}{2E_r} A^{-\frac{1}{2}}. \tag{7.8}$$

Equation (7.8) means that the relationship between S^{-1} and $A^{-\frac{1}{2}}$ is linear if Young's modulus of the sample is constant, independent of the indentation depth. Young's modulus E_s of the sample can be determined by using Eqs. (7.6) and (7.8) from the load-depth curves measured at various maximum loads after the calibration of the indenter shape. First, S^{-1} is plotted versus $A^{-\frac{1}{2}}$, and then the reduced elastic modulus E_r is obtained from the slope $\left(\frac{\sqrt{\pi}}{2E_r}\right)$ of the linear relationship between S^{-1} and $A^{-\frac{1}{2}}$ (see Fig. 7.11 for a Nb-doped TiO_2 (001) crystal). Consequently, Young's modulus of the sample can be determined from Eq. (7.6) if Poisson's ratio of the sample is known.

Alternatively, the reduced elastic modulus E_r can be obtained by analyzing the initial (elastic) portion of the load-depth curve based on the Hertzian theory [5], which predicts the elastic behavior of an indenter tip with a hemispherical end for a flat sample in the absence of frictional and adhesive interactions by using continuum elasticity. The Hertzian behavior is expressed by [6, 7]:

$$L = \frac{4}{3} E_r R^{\frac{1}{2}} h^{\frac{3}{2}}, \tag{7.9}$$

where R is the radius of an indenter tip, L is the load, and h is the indentation depth. Figure 7.6 shows schematically a load-depth curve (solid line) and the Hertzian behavior (dotted line). The dotted line in Fig. 7.6 represents Eq. (7.9). The deviation point of the load-depth curve from the Hertzian behavior is a plastic threshold at which the plastic deformation is initiated. If R is known, E_r can be determined by fitting the elastic portion of the load-depth curve with the Hertzian behavior. If R is unknown, at first, the load-depth curve for a standard sample with known E_s has to be measured and then R is obtained by fitting the elastic portion of the load-depth curve with the Hertzian behavior.

The deviation from the Hertzian behavior in Fig. 7.6 identifies the onset of plastic deformation. In some cases, this deviation is slight, but still feasible to identify, while in many cases, as shown in Fig. 7.7, an abrupt discontinuity appears in a load-depth curve, which identifies the onset of dislocation activity. This abrupt discontinuity is named "pop-in" or "excursion". The dislocation activity continues until the load rises again. The mean applied stress $\overline{\sigma}_{\mathrm{p}}$ normal to the sample surface at the initiation of dislocation activity may be expressed as follows [6]:

$$\overline{\sigma}_{\mathrm{p}} = \frac{L_{\mathrm{p}}}{\pi R h_{\mathrm{p}}}, \tag{7.10}$$

where the subscript p indicates the value at the plastic threshold. The plastic deformation results from shear stresses. It is necessary to estimate the shear stresses beneath the indenter in order to obtain a critical shear stress at which dislocations are activated. For a Hertzian stress distribution for a material with Poisson's ratio of $\nu = 0.3$, the maximum shear stress arises along the axis of symmetry at a depth of 0.48 a, where a is the radius of the contact area [5]. At this depth, the shear stress τ is given by [5]:

$$\tau = 0.31\left(\frac{3L}{2\pi R h}\right). \tag{7.11}$$

Consequently, the maximum shear stress τ_{p} at the plastic threshold load L_{p} can be estimated from the following relationship [6, 7]:

$$\tau_{\mathrm{p}} = 0.465\left(\frac{L_{\mathrm{p}}}{\pi R h_{\mathrm{p}}}\right) = 0.465\overline{\sigma}_{\mathrm{p}}, \tag{7.12}$$

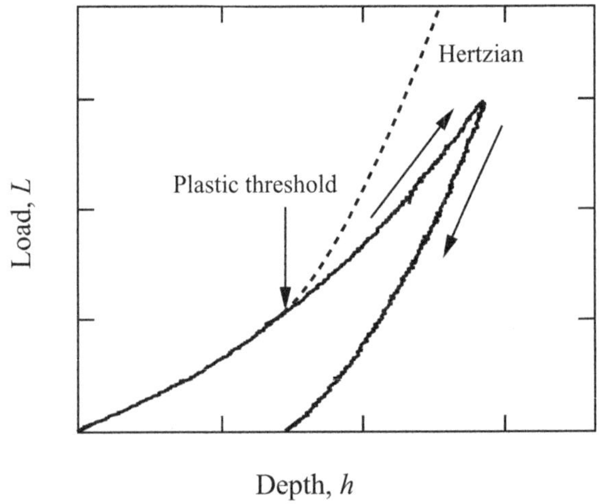

Fig. 7.6 Schematic illustration of a load-depth curve (solid line) and the Hertzian behavior (dotted line)

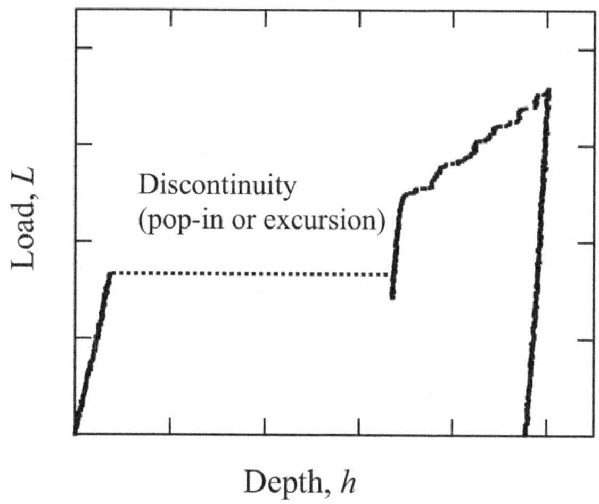

Fig. 7.7 Schematic illustration of a load-depth curve with an abrupt discontinuity which corresponds to the onset of dislocation activity

where 0.465 is the fraction of $\overline{\sigma}_{\mathrm{p}}$.

7.3 Nano-Mechanical Properties of Solid Surfaces Obtained by Nano-Indentation in Air

7.3.1 Single Crystal Gold Surfaces

There have been many studies [e.g., 8] of nano-indentation on various metal surfaces in air to evaluate the nano-mechanical properties of the metal surfaces. The load-depth curves measured by nano-indentation are often influenced by air-formed oxide films on metal surfaces [9–11] because the indentation depth is in the same order of magnitude as the thickness of air-formed oxide films (1–10 nm). If nano-indentation is performed on a bare metal surface free from an air-formed film, the nano-mechanical properties of the metal surface itself can be evaluated from the measured load-depth curve. In the present subsection, we focus on a single-crystalline gold surface as representative of a bare metal surface without an air-formed film.

There have been several studies of nano-indentation on single crystal Au surfaces [6, 7, 12–14]. For the nano-indentation study [6] of the Au (111), (001), and (110) surfaces made by using a parabolic tungsten tip, the sample surfaces were coated with a self-assembled monolayer (SAM) of hexadecane-thiol to avoid strong adhesive interaction between the Au surface and W tip prior to the indentation. The load-depth curve was not influenced by the SAM except for slight repulsive force near the contact point. The reduced moduli E_{r} and Au indentation moduli E_{Au} for flat and defect-free regions of the Au surfaces were determined by fitting the elastic portion of the loading curve with the Hertzian behavior in Eq. (7.9). The tungsten indenters with radii of

70–175 nm were employed for the measurements of the load-depth curves for two sets of the (111) / (001) and (111) / (110) surfaces. The values of $E_W = 411$ GPa, $\nu_{Au(111)} = 0.44$, $\nu_{Au(001)} = 0.46$, and $\nu_{Au(110)} = 0.45$ were used for determination of E_{Au}. The value of $E_{Au(111)}$ thus determined is 78–85 GPa, which is higher by about 36% than that of $E_{Au(001)}$[6]. On the other hand, there was no statistically significant difference between $E_{Au(111)}$ and $E_{Au(110)}$.

The indentation modulus (78–85 GPa) of the Au (111) surface is close to Young's modulus (78.7 GPa) of a polycrystalline Au [15] and that (81.3 GPa) of a single crystal Au (111) obtained from its elastic compliance data (see Sect. 3.2.3 of Chap. 3, and Eqs. (4.18) ands (4.19) in Sect. 4.3.1 of Chap. 4). The effect of crystal orientation on the indentation modulus of Au determined by using the parabolic tungsten indenter [6] is greater than the theoretical results [8] of the indentation modulus calculated by supposing the use of a rigid, flat, triangular indenter despite the same orientation dependence of the indentation modulus. The difference in magnitude of the orientation effect may be ascribed to the difference in shape and size between the indenters used for experiments [6].

Furthermore, the mean applied stresses $\overline{\sigma}_p$ of the Au (111), (001), and (110) surfaces were calculated by using Eq. (7.10) at the first deviation of the loading curve from the Hertzian behavior or the onset point of discontinuity in the loading curve. The calculated values of $\overline{\sigma}_p$ are 7.3 GPa for the (111) orientation, 5.5 GPa for the (001) orientation, and 7.8 GPa for the (110) orientation. In metals with face-centered cubic (fcc) structure such as Au, slip arises on closed-packed $\{111\}$ planes in $\langle 110 \rangle$ directions. Assuming an isotropic Hertzian stress distribution, the maximum shear stresses on $\{111\}\langle 110 \rangle$ slip systems were resolved for each of the orientations, and the resolved shear stress τ_c on the $(1\bar{1}\bar{1})[\bar{1}0\bar{1}]$ slip system at the initial yielding point was estimated by multiplying the measured mean applied stresses $\overline{\sigma}_\mathrm{p}$ with their fractions (about half of 0.465 in Eq. (7.12)) predicted from the Hertzian stress distribution. The estimated value of τ_c on the $(1\bar{1}\bar{1})[\bar{1}0\bar{1}]$ slip system is about 1.8 GPa, irrespective of orientations. The estimated value of τ_c ranges in the same level as the results obtained by other researchers [7, 12, 16] if the measured values of $\overline{\sigma}_\mathrm{p}$ are resolved to the shear components.

Nano-indentation of single crystal Au (100) thin films on NaCl substrate has been performed to explore the effect of film thickness on hardness and elastic modulus as well as on the initial plastic deformation [14]. The Au (100) thin films in the range of 32–858 nm were deposited on cleaved and polished (100) oriented NaCl single-crystalline substrates by magnetron sputtering. At first, the load-depth curves with partial unloading cycles were measured for the Au (100) thin films on NaCl substrate by using a Berkovich tip with a tip radius of 200 nm as exemplified in Fig. 7.8 [14]. The values of reduced indentation modulus E_r and hardness H were determined as a function of indentation depth from the partial unloading curves by applying the Oliver–Pharr method (see Eqs. (7.1)–(7.8)). Both the values of E_r and H decrease with increasing indentation depth, independent of the film thickness. As the indentation depth approaches the Au/NaCl interface, the value of E_r attains to that ($E_r = 43$ GPa) of the NaCl substrate. The values of E_r, E_s, and H obtained from the partial unloading curves at indentation depths of 10 and 20% of the film thickness

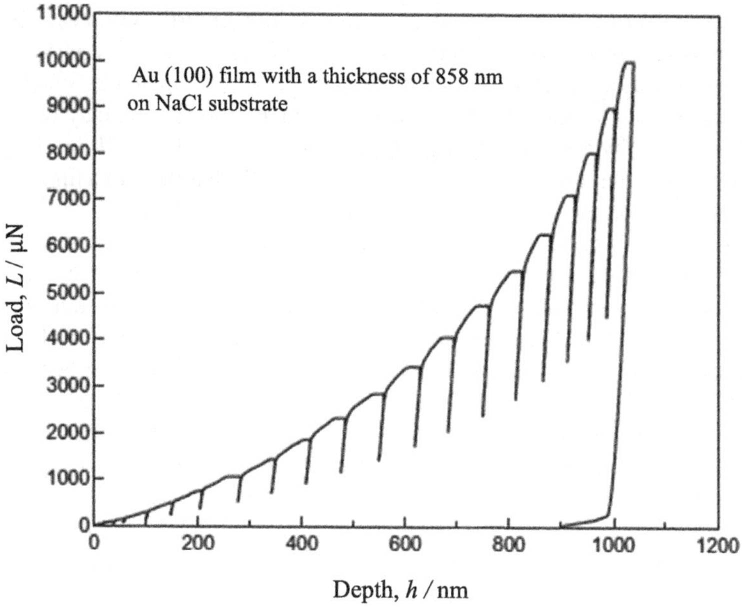

Fig. 7.8 Load-depth curve with partial unloading segments of 20 cycles for the Au (100) film (858 nm) on NaCl substrate [14]. The segment times were kept constant at 1 s for the loading segment, 0.5 s for the hold segment, and 1.5 s for the unloading segment, respectively, which leads to a total duration of 60 s per indentation. Reprinted from [14], Copyright 2008, Acta Materialia Inc, with permission from Elsevier

are listed in the second, third, and fifth columns of Table 7.1 [14], respectively. The values in the parentheses correspond to those at the indentation depth of 20% of the film thickness. The values of $\nu_{Au} = 0.44$ and $\nu_{NaCl} = 0.25$ were employed for the determination of E_s from E_r. The values of E_r, E_s, and H take maxima at a film thickness of 207 nm.

The single indentation experiments in load-control mode with constant loading and unloading rates of 30 μN s^{-1} have been also performed by using a Berkovich tip

Table 7.1 Values of reduced indentation modulus E_r, indentation modulus E_s, and hardness H for the Au (100) films obtained from partial unloading experiments at indentation depths of 10% and (20%) of the film thickness d_f in addition to the values of reduced indentation modulus $E_{r,\,Hertz}$ obtained from the Hertz fit of single indentation experiments [14]. Reprinted from [14], Copyright 2008, Acta Materialia Inc, with permission from Elsevier

d_f/nm	E_r/GPa	E_s/GPa	$E_{r,\,Hertz}$/GPa	H/GPa
134	85 (70)	74 (60)	54 ± 9	2.50 (1.70)
207	111(87)	99 (76)	73 ± 6	2.79 (1.81)
429	103 (86)	91 (75)	78 ± 20	1.76 (1.25)
858	79 (68)	68 (58)	86 ± 12	1.07 (0.71)

with a tip radius of 700 nm. The values of $E_{r,Hertz}$ in the fourth column of Table 7.1 [14] were determined from the Hertz fit (see Eq. (7.9)) of the elastic portion in the measured loading curves. In contrast to E_r and E_s, $E_{r,Hertz}$ increases steadily with increasing film thickness. The difference between E_r and $E_{r,Hertz}$ may result from the difference in indentation depth (5–10 nm for $E_{r,Hertz}$ and 13–170 nm for E_r) in addition to the difference in experimental procedure. The hardness of a film/substrate system depends on the substrate as well as indentation depth [14]. The increase in H from 1.07 (0.71) GPa to 2.79 (1.81) GPa with decreasing film thickness from 858 to 207 nm in Table 7.1 may be explained in terms of the indentation size effect (ISE) [17] and of the constraint in the dislocation motion at the film/substrate interface [18–20]. On the other hand, the decrease in H at a film thickness of 134 nm is ascribed to an additional plastic deformation in the NaCl substrate because of a distinct displacement excursion observed for a film thickness of 134 nm at a load of about 160 μN [14], corresponding to dislocation burst in the substrate. According to the nano-indentation study [12] of the single crystal Au (100) surface achieved by using a Berkovich tip with a 205 nm radius, the value of H increases from 0.6 to 2 GPa as the indentation depth decreases below 50 nm. Similarly, the ISE study [21] of face-centered cubic (fcc) single crystal metals indicated that the value of H for the single crystal Au (100) surface increases from 1.2 to 1.8 GPa as the indentation depth decreases below 50 nm.

7.3.2 Metal Oxide Surfaces

We discuss the nano-mechanical properties of single crystal TiO_2 and magnetite surfaces as typical results of nano-indentation on bulk metal oxide surfaces. A Nb (0.05 wt%)-doped rutile type of TiO_2 (001) crystal wafer (a diameter of 10 mm and a thickness of 3 mm) with an epi-polished surface was prepared for nano-indentation. The separate single indentations on the different surface positions of the TiO_2 (001) crystal wafer were achieved 20 times at each maximum load (up to $L_{max} = 3000\ \mu N$) by using a Berkovich tip, and the measured load-depth curves were averaged at each maximum load. Figure 7.9 shows the averaged load-depth curves in the depth range less than 100 nm. The hardness values H determined from the averaged load-depth curves by the Oliver–Pharr method [1, 2] are plotted versus the contact depth h_c in Fig. 7.10. The constant value of $H = 14 \pm 1$ GPa is obtained in the contact depth range of $h_c = 20 - 100$ nm for the TiO_2 (001) surface except for $H \approx 10$ GPa at $h_c \approx 10$ nm. Figure 7.11 shows the relationship between S^{-1} and $A^{-\frac{1}{2}}$. The reduced indentation modulus $E_r = 316$ GPa is obtained from the slope (2.80×10^{-12} Pa^{-1}) of the linear relationship in Fig. 7.11. If Poisson's ratio $\nu_{TiO_2} = 0.25$ is employed [22], the indentation modulus, i.e., Young's modulus $E_s = 409$ GPa is obtained for the TiO_2 (001) surface. The values of $H = 14 \pm 1$ GPa and $E_s = 409$ GPa obtained by nano-indention for the TiO_2 (001) surface are significantly higher than the micro-hardness of $H = 6 - 8$ GPa at a maximum load of $L_{max} = 0.5 - 1.0$ N and Young's modulus $E_s = 88.2$ GPa for a polycrystalline TiO_2 [23].

Nano-indentation on a single crystal magnetite (100) surface has been performed by using a cube-corner tip [24, 25]. The single crystal magnetite (100) specimen was cut from a natural octahedron crystal of magnetite and then mechanically polished to gain an optically flat surface. Figure 7.12 shows the relationship between hardness H and contact depth h_c obtained from the load-depth curves for the magnetite (100) surface [25]. The average value of H increases from 7 to 10.2 GPa with increasing h_c from 10 to 25 nm, keeps a maximum (10.5 GPa) at $h_c \approx 50$ nm, and then decreases slightly with increasing $h_c > 50$ nm. The value of $H = 10.3 \pm 0.2$ GPa in the h_c range of 25 nm to 150 nm for the magnetite (100) surface is higher than the micro-hardness value (4.7 − 7.9 GPa) at a maximum load of $L_{max} = 0.5$ N for a magnetite [23]. Young's modulus $E_s = 174$ GPa for the magnetite (100) surface is obtained by employing Poisson's ratio of $\nu_{Fe_3O_4} = 0.31$ from the linear relationship between S^{-1} and $A^{-\frac{1}{2}}$. Young's modulus and Poisson's ratio of a magnetite determined by molecular dynamic analysis [26] are $E_s = 175$ GPa and $\nu_{Fe_3O_4} = 0.37$, respectively, which are in good agreement with those obtained by nano-indentation for the magnetite (100) surface. As shown in Figs. 7.10 and 7.12, the values of H for the Nb-doped TiO_2 (001) and magnetite (100) surfaces do not change sensitively in the h_c range of 30–100 nm. On the other hand, the decreases of H in the h_c range less than 30 nm are opposite to the ISE observed in MgO [17, 27–33], which may result from the concentration gradient of doped Nb in the TiO_2 surface and from the presence of high-valency iron oxide layer such as hematite or goethite on the magnetite surface.

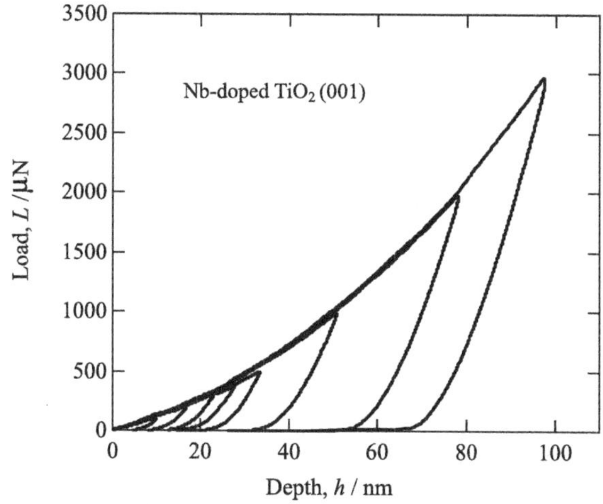

Fig. 7.9 Averaged load-depth curves in the depth range less than 100 nm for a Nb (0.05 wt%)-doped rutile type of TiO_2 (001) crystal wafer. The separate single indentations on the different surface positions of the TiO_2 (001) crystal wafer were achieved 20 times at each maximum load (up to $L_{max} = 3000$ μN) by using a Berkovich indenter and then the measured load-depth curves were averaged at each maximum load

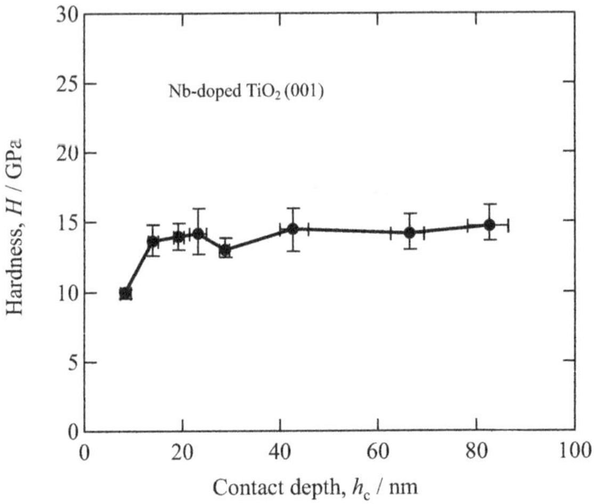

Fig. 7.10 Hardness H determined from the averaged load-depth curves (Fig. 7.9) by the Oliver–Pharr method [1, 2] as a function of contact depth h_c for the TiO_2 (001) surface

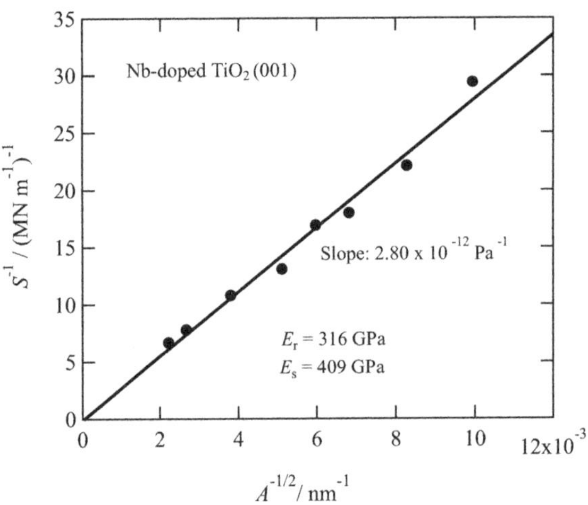

Fig. 7.11 Relationship between S^{-1} and $A^{-\frac{1}{2}}$ for the TiO_2 (001) surface. The stiffness S and projected area function A are determined from Fig. 7.9. The reduced indentation modulus E_r = 316 GPa is obtained from the slope (2.80×10^{-12} Pa^{-1}) of the linear relationship in Fig. 7.11

7.3.3 Indentation Size Effect (ISE) in MgO

The micro- and nano-indentation tests in the wide h_c range from 50 nm to 1.7 μm by using a Berkovich tip have been performed on single crystal MgO surfaces to determine the hardness as a function of indentation depth and to evaluate the ISE

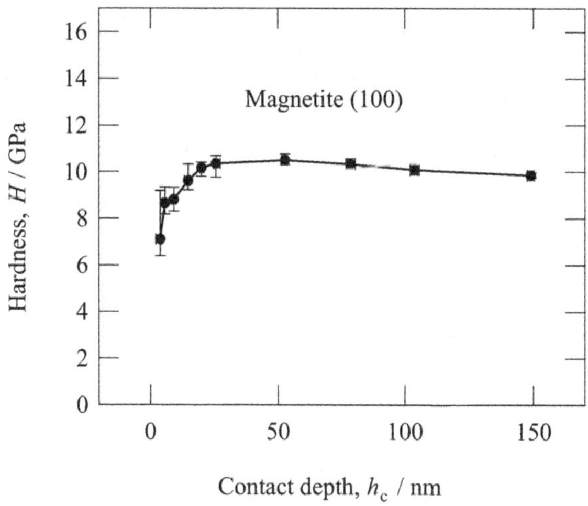

Fig. 7.12 Relationship between hardness H and contact depth h_c obtained from the load-depth curves for a single crystal magnetite (100) surface [25]. Reprinted from [25] by permission of Taylor & Francis Group LLC (CRC Press)

in MgO [27]. The cleaved and epi-polished surfaces of MgO were used for the indentation tests. The cleaved surface has facets of (100) planes with a size of about 100 μm, while the epi-polished surface has an orientation of (100)± 0.2° and a surface roughness <1 nm. The value of H decreased from 12.5 GPa to 9.5 GPa with increasing indentation depth h, and there was no observable difference between the indentation data for cleaved and epi-polished samples, indicating significant ISE for both samples. The following relationship between hardness H and indentation depth h in the micrometer depth regime has been established by Nix and Gao [28]:

$$\left(\frac{H}{H_0}\right)^2 = 1 + \left(\frac{h^*}{h}\right), \tag{7.13}$$

where H_0 is the indentation hardness in the region of $h \gg h^*$, and h^* is the characteristic scale length which depends on the properties of indented material and included angle of the indenter tip. Equation (7.13) was derived by introducing the concepts of geometrically necessary dislocations (GNDs) and of strain gradient plasticity (SGP) based on Taylor's dislocation theory. The above relationship coincided well with the micro-indentation data for polycrystalline and single crystal copper surfaces [29] as well as for single crystal silver surfaces [30].

The nano-indentation hardness data [31, 32], however, do not follow Eq. (7.13) since the relationship between H^2 and h^{-1} deviates downwards from the linearity with decreasing h in the depth range of $h < 200$ nm as shown in Fig. 7.13 for MgO [33]. The solid line in Fig. 7.13 represents the linear relationship between H^2 and h^{-1} corresponding to Eq. (7.13). The values of $H_0 = 9.27$ GPa and $h^* = 91.0$ nm are

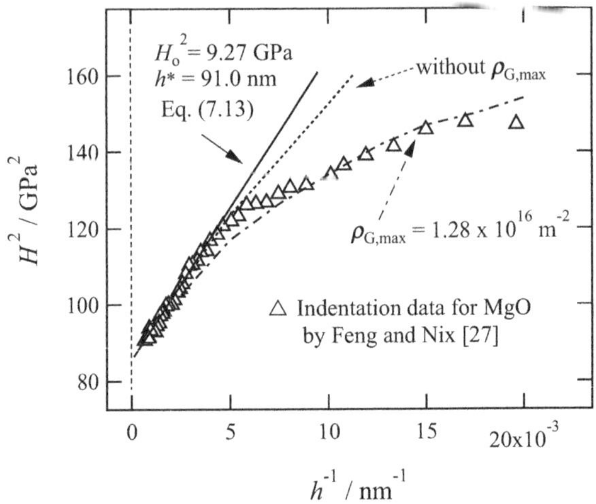

Fig. 7.13 Relationship between H^2 and h^{-1} for the indentation of MgO [33]. The symbol $\rho_{G,\,max}$ is the maximum allowable density of geometrically necessary dislocations (GND), and the triangles (△) are the indentation data of MgO obtained experimentally by Feng and Nix [27]. Reprinted from [33], Copyright 2006, with permission from Elsevier

determined from the intersection point of the solid line with the ordinate at $h^{-1} = 0$ and from the slope of the solid line. It is considered that the discrepancy between Eq. (7.13) and nano-indentation data is caused by two main factors of (I) indenter tip radius and of (II) storage volume for GNDs:

(I) Eq. (7.13) holds only for sharp pyramid indenters since the effect of indenter tip radius (around 50 nm) has not be taken into account [34, 35].

(II) Eq. (7.13) assumes that all GNDs are stored in a hemispherical zone under the contact area of indenter (see Fig. 7.14). The effective storage volume for

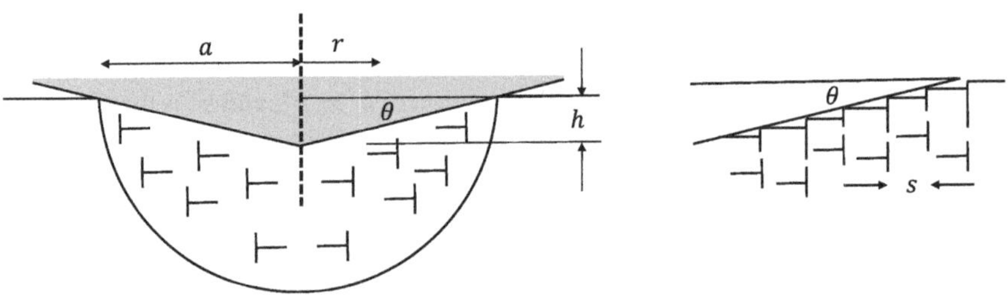

Fig. 7.14 Schematic diagram of geometrically necessary dislocations (GNDs) underneath the indenter [28, 33]. For simplicity, the indenter is assumed to be conical, and the dislocation structure is idealized as circular dislocation loops of GNDs with Burgers vector normal to the plane of the surface. The symbols shown are θ: the angle between the surfaces of the plane and indenter; a: the contact radius; r : the arbitrary radius on the contact area; h: the indentation depth; and s: the spacing between individual slip steps. Reprinted from [33], Copyright 2006, with permission from Elsevier

GNDs in nano-indentation needs to be modified because a significant number of dislocations for very small indentations spread out from the hemispherical zone [27, 31].

Huang et al. [33] developed the nano-indentation model based on a maximum allowable GND density, which is somewhat related to the effective storage volume of GNDs. The strain gradient plasticity (SGP) theory based on the Taylor dislocation model was modified to incorporate the maximum allowable GND density. In the Taylor model, the shear flow stress τ is related to the dislocation density ρ [36]:

$$\tau = \alpha\mu b\sqrt{\rho}, \tag{7.14}$$

where μ is the shear modulus, b is the magnitude of the Burgers vector, and α is an empirical coefficient of 0.3–0.5 depending on the tested materials [28, 33]. The dislocation density ρ is the sum of the density ρ_{G} for GNDs and of the density ρ_{S} for statistically stored dislocations (SSDs) [37]:

$$\rho = \rho_{\mathrm{G}} + \rho_{\mathrm{S}}. \tag{7.15}$$

The GNDs are required for the permanent shape change at the surface due to indentation, while the SSDs accumulate by random mutual trapping in the sample and contribute to the deformation resistance. In indentation, the equivalent flow stress σ is related to the shear stress τ by von Mises yield criterion [17, 33]:

$$\sigma = \sqrt{3}\tau. \tag{7.16}$$

Furthermore, H is related to σ by the Tabor factor of 3:

$$H = 3\sigma. \tag{7.17}$$

Therefore, H can be finally expressed by

$$H = 3\sqrt{3}\alpha\mu b\sqrt{\rho_G + \rho_{\mathrm{S}}}. \tag{7.18}$$

Figure 7.14 shows a schematic diagram of GNDs underneath the indenter [28, 33]. In Fig. 7.14, the SSDs are not drawn to avoid the complexity of the diagram. The calculation of ρ_G is simplified by assuming that the indenter is conical, and the indentation is accommodated by circular loops of GNDs with the Burgers vector normal to the plane of the surface. As shown in Fig. 7.14, the angle between the surfaces of the plane and indenter, the contact radius, and the indentation depth are denoted by θ, a, and h, respectively. The following relationship holds between θ, a, and h:

$$\tan\theta = \frac{h}{a} = \frac{b}{s}, \tag{7.19}$$

where b is the Burgers vector and s is the spacing between individual slip steps. In the case where a conical indenter with a cone angle of 140.6° is employed, $\theta = 19.7°$ and, thus, $\tan\theta = 0.385$. The length $\mathrm{d}\lambda$ of injected dislocation loops with radii between r and $r + dr$ can be expressed by [28]:

$$\mathrm{d}\lambda = 2\pi r\frac{\mathrm{d}r}{s} = 2\pi r\frac{h}{ab}\mathrm{d}r. \tag{7.20}$$

The total length λ of injected dislocation loops is obtained by integrating Eq. (7.20) with respect to r:

$$\lambda = \frac{\pi h a}{b}. \tag{7.21}$$

The average density ρ_{G} of GNDs is expressed by $\rho_{\mathrm{G}} = \frac{\lambda}{V}$, where $V = \frac{2}{3}\pi a^3$ is the volume of the hemisphere. Therefore, ρ_{G} is derived from Eqs. (7.19) and (7.21):

$$\rho_{\mathrm{G}} = \frac{3}{2bh}\tan^2\theta. \tag{7.22}$$

Similarly, the equivalent flow stress σ_0 in the absence of the GNDs ($\rho_{\mathrm{G}} = 0$) is derived from Eqs. (7.14)–(7.16):

$$\sigma_0 = \sqrt{3}\alpha\mu b\sqrt{\rho_{\mathrm{S}}}. \tag{7.23}$$

Furthermore, the hardness H_{o}, which arises from the SSDs in the absence of the GNDs, is derived from Eq. (7.17):

$$H_0 = 3\sqrt{3}\alpha\mu b\sqrt{\rho_{\mathrm{S}}}. \tag{7.24}$$

From the relationships among Eqs. (7.18), (7.22), and (7.24), the term of h^* in Eq. (7.13) is eventually given by

$$h^* = \frac{81}{2}b\alpha^2\tan^2\theta\left(\frac{\mu}{H_0}\right)^2. \tag{7.25}$$

The value of h^* in Eq. (7.25) is not a constant for given specimen and indenter geometry, and it depends on ρ_{S} through H_{o}. The value of $h^* = 71.3$ nm is calculated from Eq. (7.25) by using $b = 0.298$ nm, $\alpha = 0.5$, $\theta = 19.7°$, $\mu = 126$ GPa, and $H_0 = 9.27$ GPa for MgO, which is close to $h^* = 91.0$ nm obtained from the slope of the solid line in Fig. 7.13. An included angle of 142.3° for the Berkovich indenter used experiments in Fig. 7.13 is nearly equal to a cone angle of conical indenter (140.6°) employed as an ideal indenter for the derivation of Eq. (7.25). Equation (7.22) indicates that ρ_G increases with decreasing indentation depth h in the nanometer range.

However, ρ_G cannot increase infinitely because of strong repulsive forces between GNDs which impose dislocations to spread beyond the hemisphere at very small indentation depth [27, 31]. A maximum allowable GND density $\rho_{G,\,max}$ [33] was introduced to explain the nano-indentation hardness for which ρ_G can never exceed $\rho_{G,\,max}$:

$$\rho_G \leq \rho_{G,\,max}, \tag{7.26}$$

where $\rho_{G,\,max}$ is determined from experiments, and it is on the order of 10^{16} m^{-2} [33]. Furthermore, the finite element method for strain gradient plasticity (SGP) theory [33, 38–40] based on the Taylor dislocation model was employed to examine the effect of maximum allowable GND density in micro- and nano-indentation. In the finite element analysis, it is not necessary to assume GND distributions underneath the indenter, nor the storage volume for GNDs since they are automatically determined in the computations. The result obtained by the finite element analysis for MgO is represented by the dot-dash line in Fig. 7.13. The dot-dash line with a value of $\rho_{G,\,max} = 1.28 \times 10^{16} m^{-2}$ coincides well with the experimental data except for the intermediate zone between $\frac{1}{h} = 0.03$ and 0.08 nm^{-1}. The dotted line in Fig. 7.13 represents the result without consideration for $\rho_{G,\,max}$, which coincides well with the micro-indentation data $\left(\frac{1}{h} < 0.05\ nm^{-1}\right)$ but not with the nano-indentation data. In addition, the micro- and nano-indentation data for iridium as well as MgO were explained by taking the maximum allowable GND density into account [33]. The finite element analysis was also achieved for the effect of the indenter tip radius in the micro- and nano-hardness of MgO, leading to the result that the effect of the tip radius effect alone cannot explain the nano-indentation size effect without taking the maximum allowable GND density into account.

Chicot [17] proposed a new indentation parameter named " hardness length scale factor (H_{LSF})" and indicated that the H^2 versus h^{-1} curve for MgO is composed of two separate linear lines covering micro- and nano-indentation hardness data as shown in Fig. 7.15. Equation (7.13) can be rewritten by

$$H^2 = H_0^2 + \frac{H_0^2 h^*}{h}. \tag{7.27}$$

The hardness length scale factor H_{LSF} is defined by

$$H_{LSF} = H_0\sqrt{h^*}, \tag{7.28}$$

which has a unit of MPa $m^{1/2}$, and this unit is same as that of a toughness. $H_{\mathrm{LSF-micro}}$ for micro-indentation is derived from Eqs. (7.24) and (7.25):

$$H_{\mathrm{LSF-micro}} = \frac{9}{\sqrt{2}}\alpha\mu\tan\theta\sqrt{b}. \tag{7.29}$$

In nano-indentation, an indent with constant slope is not entirely consistent with the approximation of a constant strain gradient, and thus, the dislocation spacing becomes non-uniform in the plastic zone which leads to a higher density of GNDs at the center of the indent. In this case, $\rho_{\mathrm{G,\,max}}$ is given by [28]:

$$\rho_{\mathrm{G,\,max}} = \frac{1}{2bh}\tan^2\theta \tag{7.30}$$

The use of Eq. (7.30) in place of Eq. (7.22) leads to the characteristic scale length h^*_{nano} for nano-indentation:

$$h^*_{\mathrm{nano}} = \frac{27}{2}b\alpha^2\tan^2\theta\left(\frac{\mu}{H_{0,\,\mathrm{nano}}}\right)^2. \tag{7.31}$$

Consequently, $H_{\mathrm{LSF-nano}}$ for nano-indentation is given by

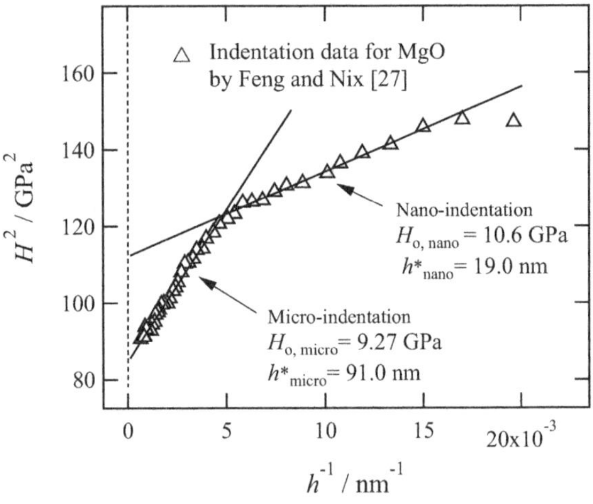

Fig. 7.15 Fitting of the two linear lines (micro- and nano-indentation regimes) with the indentation data (H^2 vs h^{-1}) of MgO achieved by introducing the hardness length scale factor H_{LSF} [17]. The triangles (△) are the indentation data of MgO obtained experimentally by Feng and Nix [27]. Reprinted from [17], Copyright 2008, with permission from Elsevier

$$H_{\text{LSF-nano}} = \frac{3\sqrt{3}}{\sqrt{2}}\alpha\mu\tan\theta\sqrt{b}, \tag{7.32}$$

which differs from $H_{\text{LSF-micro}}$ by a factor of $\sqrt{3}$. The value of $h^*_{\text{nano}} = 18.2$ nm is calculated by substituting $H_{0,\text{ nano}} = 10.6$ GPa into Eq. (7.31), which is close to that of $h^*_{\text{nano}} = 19.0$ nm obtained from the slope of the linear relation between H^2 and h^{-1} in the nano-indentation regime of Fig. 7.15. The theoretical values of $H_{\text{LSF-micro}} = 2.48$ MPa m$^{1/2}$ and $H_{\text{LSF-nano}} = 1.43$ MPa m$^{1/2}$ calculated from Eqs. (7.29) and (7.32) are consistent with the experimental values of $H_{\text{LSF-micro}} = 2.80$ MPa m$^{1/2}$ and $H_{\text{LSF-nano}} = 1.46$ MPa m$^{1/2}$.

The hardness length scale factor, $H_{\text{LSF-micro}}$ or $H_{\text{LSF-nano}}$, is proportional to $\mu\sqrt{b}$ with a factor of 2.278 α or 1.315 α as seen from Eq. (7.29) or Eq. (7.32). The experimental hardness length scale factor $H_0\sqrt{h^*}$ obtained from the indentation data of metals as well as MgO was plotted as a function of the theoretical value of $\mu\sqrt{b}$[17]. The linear relationship between $H_0\sqrt{h^*}$ and $\mu\sqrt{b}$ held in both micro- and nano-indentation regimes. The slopes were 1.17 and 0.65 for the micro- and nano-indentation data, respectively, from which the corresponding values of $\alpha = 0.51$ and 0.49 were obtained [17]. The above value of α is close to $\alpha = 0.5$ employed for the theoretical calculation of hardness parameters for MgO. Furthermore, Chicot [17] found that the following empirical relationship holds between the ratio $\chi = \frac{H_{0,\text{ micro}}}{H_{0,\text{ nano}}}$ and μ:

$$\chi^2 = \left(\frac{H_{0,\text{ micro}}}{H_{0,\text{ nano}}}\right)^2 = \left(\frac{\mu}{150}\right)^{3/2}. \tag{7.33}$$

If the shear modulus μ of the specimen is known, $H_{0,\text{micro}}$ can be calculated from $H_{0,\text{nano}}$ by using Eq. (7.33). Besides, the hardness length scale factor $H_{\text{LSF, nano}}$ obtained in nano-indentation regime allows the calculation of $H_{\text{LSF, micro}}$ and h^*_{micro} in the micro-indentation regime, that is, it is possible to construct the relationship between H^2_{micro} and h^{-1}_{micro} in micro-indentation from nano-indentation data [17].

7.3.4 Anodic Oxide Films on Metals

The nano-mechanical properties of anodic oxide films on metals have been investigated by nano-indentation [41]. A barrier-type, amorphous anodic oxide (alumina) film with a thickness of 490 ± 4 nm was formed on Al at 5 mA cm^{-2} in 0.01 M ammonium pentaborate solution for nano-indentation. The nano-indentation depth was limited to a shallow depth of 55 nm corresponding to about one-tenth of the film thickness in order to avoid the influence of the substrate Al. The values of $H = 7.0 \pm 0.7$ GPa and $E_s = 122 \pm 12$ GPa for the anodic alumina film were obtained by nano-indentation using a Berkovich indenter. Poisson's ratio $\nu = 0.234$

of the anodic alumina film was employed for the determination of E_s from the measured E_r. The value of $H = 7.0$ GPa obtained for the alumina film is significantly low as compared to $H = 25.9$ GPa for sapphire (111) [42], while it is higher than that ($H = 3 - 5$ GPa) for the porous alumina film with an average thickness of 744 nm on a glass substrate ($H = 9 - 12$ GPa) [43]. The high value of H for the alumina film relative to the porous alumina film is attributed to the absence of porosity. On the other hand, the significantly low value of H for the alumina film as compared to sapphire (111) indicates that the anodic alumina film is soft relative to crystalline forms.

The value of $E_s = 122$ GPa for the anodic alumina film is lower by a factor of 3.2 or 3.5 than $E_s = 387$ GPa for γ-Al_2O_3 [44] or $E_s = 433$ GPa for sapphire (111) [42], which is matched by the low hardness value of the anodic alumina film. It has been reported that Young's modulus of γ-Al_2O_3 decreases rapidly with increasing humidity [44]. The low Young's modulus of the anodic alumina film may result from moisture absorbed or water remained in the film during anodic oxidation. The low values of H and E_s for the anodic alumina film afford proof that the anodic alumina film is easily subjected to a plastic flow due to the evolution of compressive stress during anodic oxidation of Al, which leads to the growth of the porous alumina layer (see Sect. 6.9 of Chap. 6).

For comparison with the anodic alumina film, nano-indentation [41] was also performed in the same shallow depth by using a Berkovich indenter for the anodic oxide (tantala) film with a thickness of 408 ± 4 nm and for the mixed anodic oxide (alumina/tantala) film with a thickness of 484 ± 4 nm, which were prepared with anodic oxidation (at 5 mA cm^{-2} in ammonium pentaborate solution) of sputter-deposited tantalum and Al-50 at% Ta alloy on Al foil substrate, respectively. The values of $H = 5.3 \pm 0.5$ GPa obtained for the anodic tantala film and of $H = 6.5 \pm 0.7$ GPa obtained for the mixed anodic film are lower than that for the anodic alumina film, while the values of $E_s = 140 \pm 14$ GPa for the anodic tantala film and of $E_s = 130 \pm 13$ GPa for the mixed anodic film are higher than that for the anodic alumina film. The hardness of the mixed anodic film is closer to that of the anodic alumina film than that of the anodic tantala film, which is consistent with the aluminum-enrichment in the outer layer (thicker than the indentation depth of 55 nm) of the mixed anodic film [46]. The anodic tantala films are amorphous by X-ray diffraction, but they contain a local structure similar to β-Ta_2O_5 [46, 47]. In addition, the micro-diffraction by transmission electron microscopy (TEM) revealed the presence of micro-crystallinity in the anodic tantala films [48]. On the other hand, there is no information on the structure of the mixed oxide films.

The issue for determination of hardness and elastic modulus from the measured load-depth curves is that the Oliver–Pharr method [1, 2] does not account for pileup of material around the indenter impression as shown schematically in Fig. 7.16a. When pileup occurs, the contact area is larger than that estimated by the Oliver–Pharr method, and both hardness and elastic modulus determined by Eqs. (7.1) and (7.8) are overestimated, sometimes by as much as 50% [49]. The Oliver–Pharr method is based on an elastic contact analysis to estimate the contact depth. Since sink-in (see Fig. 7.16b) always occurs for materials deformed elastically, pile-up cannot properly

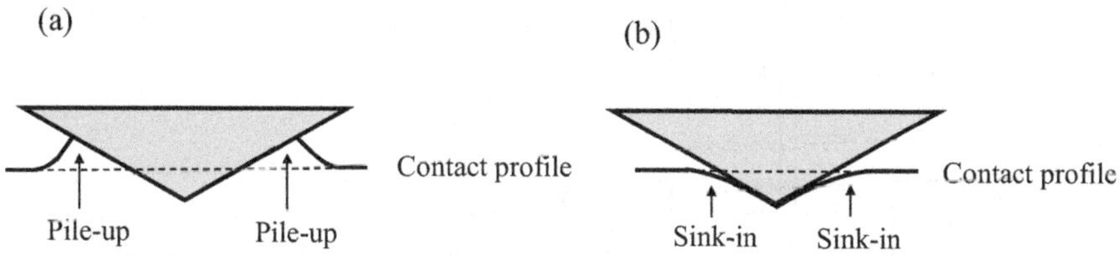

Fig. 7.16 Schematic illustration of the contact profiles. **a** pile-up, and **b** sink-in for material surface around indenter impression

be modeled. The fundamental mechanical properties of materials are associated with the ratio of reduced modulus E_r to yield stress σ_y and work-hardening behavior [2, 42, 49]. In general, materials with large $\frac{E_r}{\sigma_y}$ are subjected to a significant pile-up, while work-hardened materials have little or no capacity of pile-up. The finite element simulations [2, 42, 49] indicated that the amount of pile-up or sink-in depends on the ratio of h_f to h_{max} in load-depth curves and on the degree of work-hardening. Particularly, the noteworthy result is that in the case of $\frac{h_f}{h_{max}} < 0.7$, there is very little pile-up irrespective of the work-hardening behavior, that is, the Oliver–Pharr method provides a reasonable estimate of the contact area. The load-depth curves disclosed the values of $\frac{h_f}{h_{max}} \approx 0.63$, 0.64 and 0.74 for the anodic alumina film, mixed anodic film, and anodic tantala film, respectively [41]. Although the value of $\frac{h_f}{h_{max}}$ for the anodic tantala film exceeds slightly 0.7, the overestimation of hardness and elastic modulus is about 3% for a material without work hardening [41].

7.4 Nano-Mechanical Properties of Passive Metal Surfaces Obtained by Electrochemical Nano-Indentation

Electrochemical nano-indentation has been performed to investigate the nano-mechanical properties of surface oxide films formed on metals in electrolyte solutions. The corrosion resistivities of metals and alloys are sustained by very thin and compact surface oxide films (named "passive films") with a thickness of 1–10 nm [50]. The physicochemical properties of the passive films such as thickness, chemical composition, and semiconductive structure [50, 51] are well known, while the knowledge of their nano-mechanical properties is lacking to understand essentially the passivity of metals and alloys.

Figure 7.17 shows schematically an electrochemical cell designed for electrochemical nano-indentation [52]. A miniature cell is made from polytetrafluoroethylene (PTFE), and a working electrode is a metal disk plate with a diameter of 10 mm, which is attached to the bottom of the cell by using O-ring, screw, and stainless steel-disk plate. Platinum wire and ring are employed as reference electrode and counter electrode, respectively. The potential of the Pt wire is always measured with an Ag/AgCl reference electrode for calibration before and after every experiment.

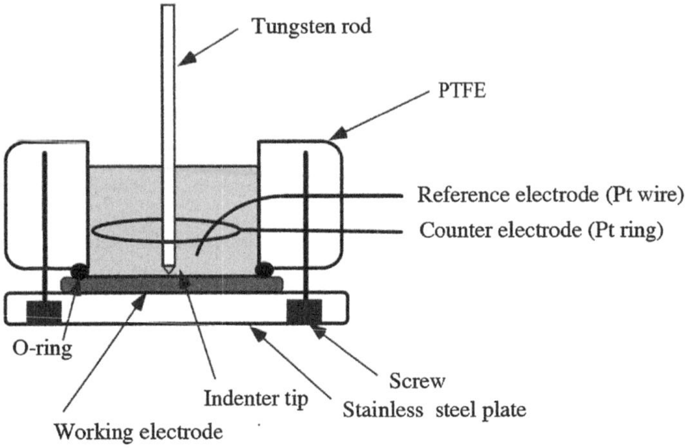

Fig. 7.17 Schematic structure of an electrochemical cell designed for electrochemical nano-indentation [52]. Reprinted from [52], Copyright 2002, with permission from Elsevier

The working, counter, and reference electrodes are connected through lead wires to a potentiostat by which the working electrode is electrochemically controlled in electrolyte solution.

7.4.1 Passive Single Crystal Fe (100) and (110) Surfaces in Solution

Single crystal Fe (100) and (110) disk plates have been used for electrochemical nano-indentation [24, 53]. The iron specimen was mechanically polished with alumina abrasives and then electro-polished with a constant current density of 32 mA cm^{-2} in a mixture of 70% $HClO_4$ and glacial CH_3COOH (1:20) at 17 °C. The electrolyte solution employed is pH 8.4 borate solution, well known as an electrolyte which stabilizes the passivity of ferrous metals [54]. Figure 7.18 shows the potentiodynamic polarization curves of the Fe (100) and (110) surfaces measured at a potential scan rate of 100 mV min^{-1} in pH 8.4 borate solution [24]. The anodic current peak of the Fe (110) surface in the active dissolution region is three times as much as that of the Fe (100) surface. By contrast, there are no significant differences between Fe (100) and (110) surfaces in the passive region. The high anodic current density of the Fe (110) surface may result from its high surface atomic density which is 1.4 times as much as that of Fe (100) surface.

The electrochemical nano-indentation experiments of the Fe (100) and (110) surfaces were performed by using a cube-corner indenter (see Fig. 7.1b) with a tip radius less than 100 nm. The Fe specimens were cathodically reduced with a constant current density of −30 μA cm^{-2} in pH 8.4 borate solution to remove an air-formed film. After the cathodic reduction, the Fe specimens were anodically polarized at a constant potential in the passive potential region between 0.0 V and 1.0 V (SHE) for 1 h in the renewed solution for electrochemical nano-indentation. The load-depth curves of the Fe specimens passivated at a constant potential were measured twenty times at a maximum load of $L_{max} = 100 - 500\ \mu N$ for each loading and unloading time of 5 s and then were averaged because of fluctuation in each load-depth curve.

Figure 7.19 shows the averaged load-depth curves of the Fe (100) and (110) surfaces passivated at 0.25 V (SHE) in pH 8.4 borate solution [24]. The indentation depth h_{max} at each maximum load for the passive Fe (110) surface is always smaller than that for the passive Fe (100) surface, indicating that the hardness of the passive Fe (110) surface is higher than that of the passive Fe (100) surface. The similar averaged load-depth curves at various maximum loads were obtained for the Fe (100) and (110) surfaces passivated at other potentials. The values of hardness H for the passive Fe (100) and Fe (110) surfaces determined from the averaged load-depth curves at $L_{max} = 400\ \mu N$ by using the Oliver–Pharr method [1, 2] are plotted as a function of applied potential in Fig. 7.20 [24]. The values of H for the passive Fe (100) and Fe (110) surfaces range from 2.9 GPa to 3.3 GPa, and they tend to increase linearly with increasing potential. The slope of the linear relationship between H and E is about 0.2 GPa V^{-1}, irrespective of the Fe crystal orientation, although the value of H for the passive Fe (110) surface is higher by about 10% than that for the passive Fe (100) surface.

An ellipsometrical study [54] indicated that the thickness of passive film formed on polycrystalline iron for 1 h at a constant potential in pH 8.4 borate solution increases linearly with increasing potential. The dashed line in Fig. 7.20 represents the linear relationship between thickness of passive film d_f and potential E. The

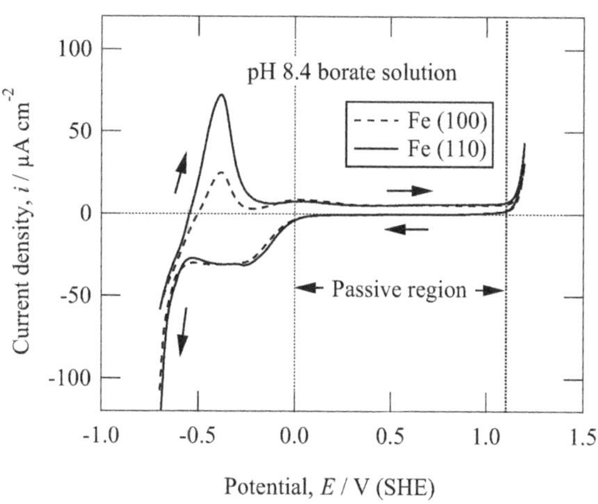

Fig. 7.18 Potentiodynamic polarization curves of the Fe (100) and (110) surfaces measured at a potential scan rate of 100 mV min^{-1} in pH 8.4 borate solution [24]. Reprinted from [24], Copyright 2001, with permission from Elsevier

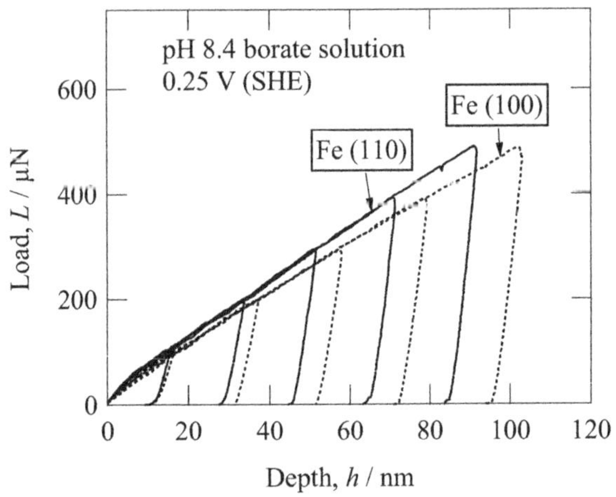

Fig. 7.19 Averaged load-depth curves of the Fe (100) and (110) surfaces passivated at 0.25 V (SHE) in pH 8.4 borate solution [24]. The load-depth curves measured twenty times at a maximum load of $L_{max} = 100–500$ μN for each loading and unloading time of 5 s were averaged because of fluctuation in each load-depth curve. Reprinted from [24], Copyright 2001, with permission from Elsevier

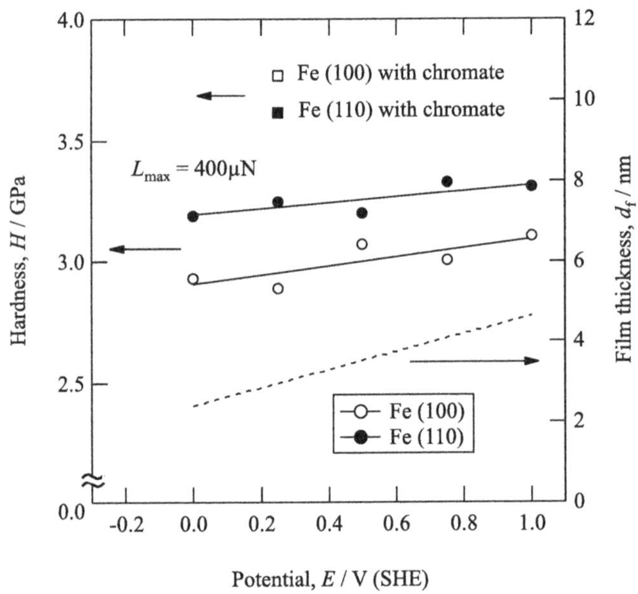

Fig. 7.20 Values of hardness H for the passive Fe (100) and Fe (110) surfaces determined from the averaged load-depth curves at $L_{max} = 400$ μN as a function of applied potential [24]. The dashed line represents the linear relationship between thickness of passive film d_f and potential E[54]. The square marks (□ and ■) represent the values of $H = 3.75$ and 3.62 GPa determined from the averaged load-depth curves at $L_{max} = 400$ μN, respectively, for the Fe (100) and Fe (110) surfaces passivated at 0.25 V (SHE) in pH 8.4 borate solution after the chromate treatment. Reprinted from [24], Copyright 2001, with permission from Elsevier

slope of the dashed line is about 2.5 nm V^{-1}. The values of H for bare Fe (110) and Fe (100) surfaces without passive film may be estimated to be 3.0 and 2.7 GPa, respectively, from the potential dependences of hardness (0.2 GPa V^{-1}) and film thickness (2.5 nm V^{-1}).

Metals with a body-centered cubic structure (bcc) such as α-Fe have slip planes of $\{110\}$ and $\{112\}$ with a slip direction of $\langle 111 \rangle$ [15]. The difference in H between bare Fe (100) and Fe (110) surfaces may result from the difference in resolved shear stress on the slip system $\{110\}\langle 111 \rangle$ or $\{112\}\langle 111 \rangle$ at the initial yielding point for plastic deformation. The contact depth h_c is in the range of 65–90 nm at $L_{max} = 400$ μN. Consequently, the ratio of d_f to h_c is only in the range from 0.03 to 0.05. The linear increase in H with potential for both passive Fe (100) and Fe (110) surfaces in response to the linear increase in d_f with potential is ascribed to the contribution of the passive film with a high value of H despite very small values of $\frac{d_f}{h_c}$. It is known that the structures of passive films on iron single crystals (001) and ($1\bar{1}0$) are similar to that of magnetite [55, 56]. The value of H obtained by nano-indentation for the magnetite single crystal (100) surface [24] (see Fig. 7.12) is 10.2 GPa which is about 3.4 times as much as those of the passive Fe (100) and (110) surfaces. The value of E_s (Young's modulus) obtained for the passive Fe (100) and (110) surfaces is 175 ± 30 GPa [24], which is close to that ($E_s = 174$ GPa) obtained by nano-indentation for the magnetite single crystal (100) surface, but it is lower than that ($E_s = 208.2$ GPa) for bulk α-Fe [15].

7.4.2 Effect of Chromate Treatment on Nano-Mechanical Properties of Passive Fe Surfaces

It is well known [58, 59] that the corrosion resistivities of metals such as Al, Fe, and steels are significantly improved by a chromate treatment (i.e. by immersion of metals into chromic acid or dichromate solution). The chromate treatment is not used nowadays because hexavalent chromium is toxic, and thus, highly regulated. Nevertheless, it is worthy from the scientific viewpoint to investigate the effect of the chromate on nano-mechanical properties of the passive metals which are directly associated with the corrosion resistivities. For this purpose, electrochemical nano-indentation experiments have been performed to the passive Fe (100) and (110) surfaces without and with chromate treatment [24, 52]. The chromate treatment was made with natural immersion of the Fe specimens in 5×10^{-2} M $K_2Cr_2O_7$ solution for 24 h.

The solid lines in Fig. 7.21 represent the averaged load-depth curves for the Fe (100) surface passivated at 0.25 V (SHE) for 1 h in pH 8.4 borate solution after the chromate treatment [24]. For comparison, the averaged load-depth curves for the Fe (100) surface passivated without the chromate treatment are represented by the dashed lines in Fig. 7.21. It is clear that the hardness of the passive Fe (100) surface is increased by the chromate treatment. Furthermore, the slope of the unloading

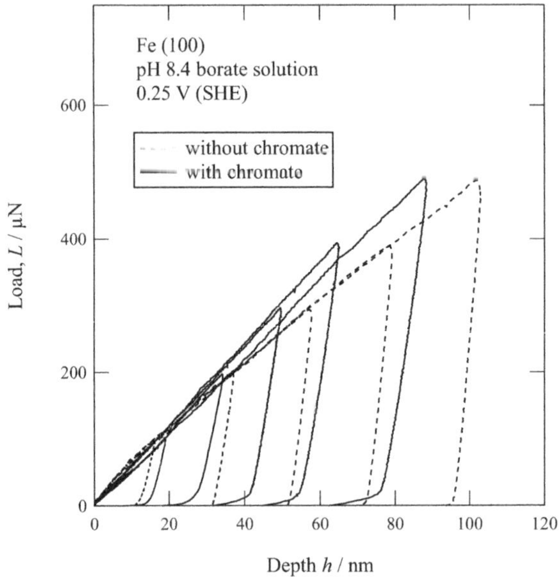

Fig. 7.21 Averaged load-depth curves (solid lines) for the Fe (100) surface passivated at 0.25 V (SHE) for 1 h in pH 8.4 borate solution after the chromate treatment [24]. For comparison, the load-depth curves for the Fe (100) surface passivated without the chromate treatment are represented by the dotted lines. The chromate treatment was made with natural immersion of the Fe specimens in 5×10^{-2} M $K_2Cr_2O_7$ solution for 24 h. Reprinted from [24], Copyright 2001, with permission from Elsevier

curve at each L_{max}, i.e. the stiffness S of the passive Fe (100) surface is somewhat reduced by the chromate treatment. The reduction in stiffness S corresponds to the decrease in Young's modulus E_s. The similar effects of the chromate treatment on the load-depth curves were observed for the passive Fe (110) surface [24]. The square marks (□ and ■) in Fig. 7.20 represent the values of $H = 3.75$ GPa and 3.62 GPa determined from the load-depth curves at $L_{max} = 400$ μN, respectively, for the Fe (100) and Fe (110) surfaces passivated at 0.25 V (SHE) in pH 8.4 borate solution after the chromate treatment [24]. The effect of the chromate treatment on hardness of the passive Fe (100) surface is more significant than that of the passive Fe (110) surface. Young's moduli $E_s = 134$ GPa and 112 GPa determined, respectively, for the passive Fe (100) and (110) surfaces after the chromate treatment were lower than those ($E_s = 175 \pm 30$ GPa) for the passive Fe (100) and (110) surfaces without the chromate treatment [24]. Both the increase in H and the reduction in E_s of the passive Fe (100) and (110) surfaces due to the chromate treatment may be associated with the promotion of passivation due to incorporation of trivalent chromium (Cr^{3+}) in passive film. Hexavalent chromium (Cr^{6+}) in solution is one of the strong oxidative agents which is known as a passivator [57, 58].

The following coupled redox reactions on the iron surface take place during the chromate treatment:

$$Cr_2O_7^{2-} + 8H^+ + 6e^- \rightarrow Cr_2O_3 + 4H_2O, \quad (7.34)$$

and

$$2Fe + 3H_2O \rightarrow Fe_2O_3 + 6H^+ + 6e^-, \quad (7.35)$$

where e^- is the electron which participates in cathodic reaction of Eq. (7.34) and in anodic reaction of Eq. (7.35). As a result of the above coupled reactions, trivalent chromium is incorporated in the passive film. The significant amount of chromium in the passive film on the polycrystalline iron after the chromate treatment has been confirmed by Auger electron spectroscopy (AES) [59, 60]. It is well known that the high corrosion resistivities of alloys containing Cr component are caused by the Cr enrichment in the passive films [61]. As shown in the load-depth curves in Figs. 7.19 and 7.21, the values of $\frac{h_f}{h_{max}}$ exceed 0.8, which may arise the issue of pile-up. Iron and steels are work-hardening materials in contrast to soft materials such as copper and aluminum. In the case where the Fe (100) and (110) specimens are work hardened during the loading required for 5 s up to each L_{max}, the influence of the pile-up on H and E_s should be very small [2, 42, 49].

As seen from the load-depth curve schematically drawn in Fig. 7.22, total work W_t done with electrochemical nano-indentation consists of two terms: work W_p due to plastic deformation and work W_e due to elastic deformation which correspond to the

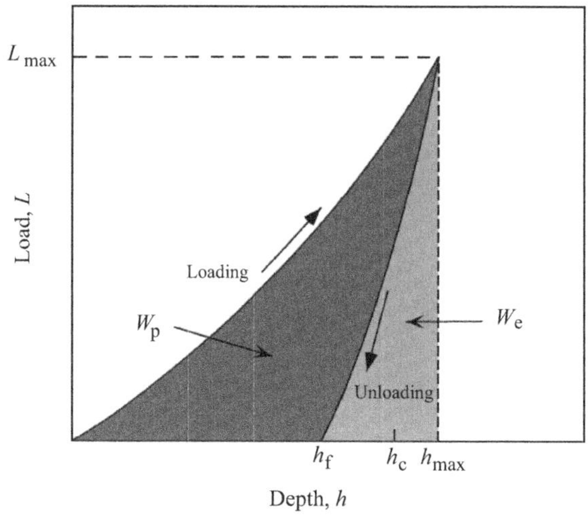

Fig. 7.22 Total work W_t done during electrochemical nano-indentation which consists of two terms: work W_p due to plastic deformation and work W_e due to elastic deformation ($W_t = W_p + W_e$) [52]. W_p and W_e correspond to the areas hatched with dark- and light-gray colors, respectively, in the schematic load-depth curve. Reprinted from [52], Copyright 2002, with permission from Elsevier

areas hatched with dark- and light-gray colors, respectively, in the schematic load-depth curve [52]. From the finite element simulations [62–65] of many elastic–plastic materials with different work-hardening behaviors for pile-up during indentation, it is deduced that the work ratio of W_e to W_t is a unique function of the ratio of reduced modulus E_r to hardness H, independent of the work-hardening behavior. Although the presented results are only in graphical form, the following relationship [2] can be approximated:

$$\frac{W_e}{W_t} \cong 5\frac{H}{E_r}. \tag{7.36}$$

Figure 7.23 shows the values of hardness H and work ratio $\frac{W_e}{W_t}$ as a function of contact depth h_c determined from the load-depth curves for the Fe (100) surfaces passivated at 0.25 V (SHE) with and without the chromate treatment [52]. In the h_c range from 40 to 60 nm, the passive Fe (100) surface with the chromate treatment has the values of $H \approx 3.7$ GPa and $\frac{W_e}{W_t} \approx 0.12$, while that without the chromate treatment has the values of $H \approx 3.1$ GPa and $\frac{W_e}{W_t} \approx 0.07$. In addition, the reduced

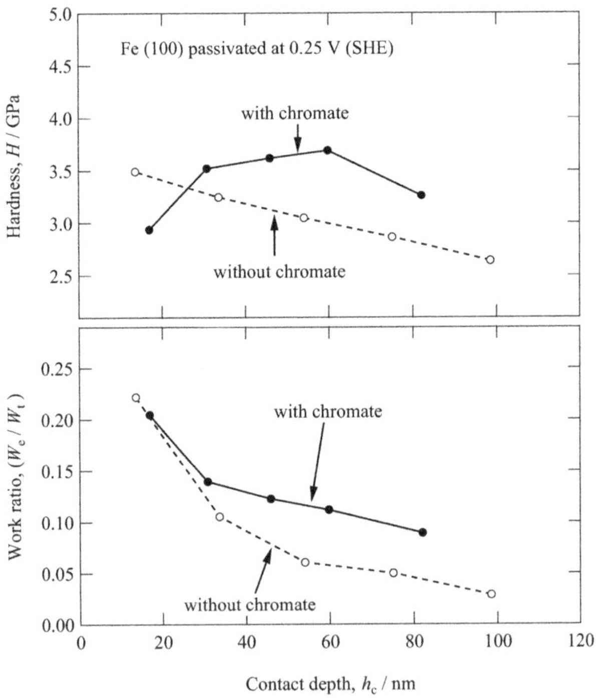

Fig. 7.23 Effects of the chromate treatment on hardness H and work ratio $\frac{W_e}{W_t}$ for the passive Fe (100) surface at 0.25 V (SHE) in pH 8.4 borate solution [52]. Reprinted from [52], Copyright 2002, with permission from Elsevier

moduli determined by using Eq. (7.8) from the measured load-depth curves of the passive Fe (100) surfaces with and without the chromate treatment were $E_r = 145$ GPa and180 GPa, respectively [24]. Consequently, the values of $5\frac{H}{E_r} \approx 0.13$ and 0.08 are calculated, respectively, for the passive Fe (100) surface with and without the chromate treatment, which are in good agreement with $\frac{W_e}{W_t} \approx 0.12$ and 0.07, respectively, obtained experimentally from Fig. 7.23 [52]. The validity of Eq. (7.36) is proved for both the passive Fe (100) surfaces with and without the chromate treatment. In Fig. 7.23, the contact depth range where the significant effect of the chromate treatment emerges is $h_c \approx 40 \sim 60$ nm, which is much larger than the thickness $d_f \approx 3$ nm of the passive film on iron at 0.25 V (SHE) in pH 8.4 borate solution. The chromate treatment does not alter or rather reduce the thickness of the passive film on iron [60].

When the indentation depth exceeds a critical level by several times as much as the film thickness, the passive film would be locally ruptured at the very edge of the contact surface where the tensile stress becomes maximum [5]. The local rupture (breakdown) of the passive film at the very edge of the contact surface may promote the depth-displacement of the indenter. Assuming that the passive film (Cr-enriched) with the chromate treatment has a fracture toughness higher than the passive film without the chromate treatment, a critical load level of the local rupture for the passive film with the chromate treatment should be higher than that for the passive film without the chromate treatment. The increase in fracture toughness of the passive film due to the chromate treatment would bring the increase in hardness since the passive film resists the depth-displacement of the indenter as far as the load does not exceed a critical level of the local film rupture.

Furthermore, in the case where the load exceeds a critical level of the local film rupture, the rupture sites may be repaired by the formation of a new oxide at the sites. It has been reported that the Cr component enriched in the passive film promotes the repair (re-passivation) of the rupture sites [66, 67], suggesting that the repair at the rupture sites for the passive surface with the chromate treatment is faster than that for the passive surface without the chromate treatment. The fast formation of the new oxide at the rupture sites due to the chromate treatment leads to the rapid repair, which would resist the depth-displacement of the indenter to bring the increase in hardness. The fast formation of the new oxide at the rupture sites due to the chromate treatment may influence a shape of the unloading curve as seen from Fig. 7.21. Particularly, the tailing at the latter stage of the unloading curve in addition to the decrease in stiffness S (corresponding to the initial slope of the unloading curve) seems to be brought by the raid repair (re-passivation) at the rupture sites due to the chromate treatment. Therefore, it is deduced that the increase in hardness and the decrease in reduced elastic modulus (i.e., the increase in work ratio $\frac{W_e}{W_t}$) of the passive Fe (100) surfaces due to the chromate treatment result from the high fracture toughness and rapid repair of the passive films. Further experimental evidence, however, is needed to confirm the above speculation.

References

1. Oliver WC, Pharr GM (1992) J Mater Res 7:1564–1583
2. Oliver WC, Pharr GM (2004) J Mater Res 19:3–20
3. Bhushan B (1995) Handbook of micro/nano tribology. CRC Press, New York, pp 321–396
4. Sneddon IN (1965) Int J Engng Sci 3:47–57
5. Johnson KL (1985) Contact mechanics. Cambridge Univ Press, Cambridge
6. Kiely JD, Houston JE (1998) Phys Rev B 57:12588–12594
7. Michalske TA, Houston JE (1998) Acta Mater 46:391–396
8. Vlassak JJ, Nix WD (1994) J Mech Phys Solids 42:1223–1245
9. Venkataraman SK, Kohlstedt DL, Gerberich WW (1993) J Mater Res 8:685–687
10. Gerberich WW, Venkataraman SK, Huang H, Harvey SE, Kohlstedt DL (1995) Acta Metall Mater 4:1569–1576
11. Bahr DF, Kramer DE, Gerberich WW (1998) Acta Mater 46:3605–3617
12. Corcoran SG, Colton RJ, Lilleodden ET, Gerberich WW (1997) Phys Rev B 55:R16057–R16060
13. Carrasco E, González MA, Rodríguez de la Fuente O, Rojo JM (2004) Surf Sci 572:467–475
14. Dietiker M, Nyilas RD, Solenthaler C, Apolenak R (2008) Acta Mater 56:3887–3899
15. Reid CN (1973) Deformation geometry for materials scientists. Pergamon Press, London
16. Stalder A, Dürig U (1996) Appl Phys Lett 68:637–639
17. Chicot D (2009) Mater Sci Eng A 499:454–461
18. Freund LB (1990) J Appl Phys 68:2073–2080
19. Ovecoglu ML, Doerner MF, Nix WD (1987) Acta Metall 35:2947–2957
20. Nix WD (1997) Mater Sci Eng A 234–236:37–44
21. Zong Z, Soboyejo W (2005) Mater Sci Eng A 404:281–290
22. Nelson JC, Oriani RA (1993) Corrs Sci 34:307–326
23. Samsonov GV (1973) The oxide handbook. IFI/Plenum, New York
24. Seo M, Chiba M (2001) Electrochim Acta 47:319–325
25. Seo M (2006) Nano-indentation and nano-scratching techniques for evaluation of mechanical properties of surface films. In: Marcus P, Mansfeld F (eds) Analytical methods in corrosion science and engineering. CRC Press, New York
26. Chicot D, Mendoza J, Zaoui A, Louis G, Lepingle V, Roudet F, Lesage J (2011) Mater Chem Phys 129:862–870
27. Feng G, Nix WD (2004) Scripta Mater 51:599–603
28. Nix WD, Gao H (1998) J Mech Phys Soilds 46:411–425
29. McElhaney KW, Vlassak JJ, Nix WD (1998) J Mater Res 13:1300–1306
30. Ma Q, Clarke DK (1995) J Mater Res 10:853–863
31. Swadener JG, Geroge EP, Pharr GM (2002) J Mech Phys Solids 50:681–694
32. Lim YY, Chaudhri MM (1999) Philos Mag A 79:2979–3000
33. Huang Y, Zhang F, Hwang KC, Nix WD, Pharr GM, Feng G (2006) J Mech Phys Solids 54:1668–1686
34. Qu S, Huang Y, Nix WD, Jiang H, Zhang F, Hwang KC (2004) J Mater Res 129:3423–3434
35. Kim JY, Lee BW, Read DT, Kwon D (2005) Scr Mater 52:353–358
36. Taylor GI (1934) Proc R Soc London A 145:362–387
37. Ashby MF (1970) Philos Mag 21:399–424
38. Gao H, Huang Y, Nix WD, Hutchinson JW (1999) J Mech Phys Solids 47:1239–1263
39. Huang Y, Gao H, Nix WD, Hutchinson JW (2000) J Mech Phys Solids 48:99–128
40. Huang Y, Xue Z, Gao H, Nix WD, Xia ZC (2000) J Mater Res 15:1786–1796
41. Alcalá G, Skeldon P, Thompson GE, Mann AB, Habazaki H, Shimizu K (2002) Nanotech 13:451–455
42. Pharr GM (1998) Mater Sci Eng A 253:151–159
43. Hu Z, Shrestha M, Fan QH (2016) Thin Solid Films 598:131–140
44. Grosskreutz JC (1969) J Electrochem Soc 116:1232–1237

45. Habazaki H, Shimizu K, Skeldon P, Thompson GE, Wood GC (1997) Proc R Soc A 453:1593–1609
46. Vermilyea DA (1957) J Electrochem Soc 104:485–488
47. Alshina LA, Malinenko VP, Phouphanov AD, Jakoleva NM (1986) J Non-Cryst Solids 87:485-488
48. Shimizu K, Thompson GE, Wood GC, Kobayashi K (1991) Phil Mag B 63:891–899
49. Bolshakov A, Pharr GM (1998) J Mater Res 13:1049–1058
50. Sato N (1990) Corros Sci 31:1–19
51. Stimming U, Schultze JW (1979) Electrochim Acta 24:859–869
52. Chiba M, Seo M (2002) Corros Sci 44:2379–2391
53. Seo M, Chiba M, Suzuki K (1999) J Electroanal Chem 473:49–53
54. Sato N, Kudo K, Noda T (1971) Electrochim Acta 16:1909–1921
55. Toney MF, Davenport AJ, Oblonsky LJ, Ryan MP, Vitus CM (1997) Phys Rev Lett 24:4282–4285
56. Davenport AJ, Oblonsky LJ, Ryan MP, Toney MF (2000) J Electrochem Soc 147:2162–2173
57. Fontana MG, Green ND (1967) Corrosion engineering. MacGraw-Hill, New York, p 217
58. Uhlig HH, Revie RW (1985) Corrosion and corrosion control: an introduction to corrosion science and engineering. 3rd edn. John Wiley & Sons, New York, pp 263–271
59. Lumsden JB, Szklarska-Smialowska Z (1978) Corrosion 34:169–176
60. Seo M, Sato N (1989) Inhibition in the context of passivation: paper-138:1–18. Corrosion-89. NACE, Houston
61. Strehblow H-H, Marcus P (2006) X-ray photoelectron spectroscopy in corrosion research. In: Marcus P, Mansfeld F (eds) Analytical methods in corrosion science and engineering. CRC Press, New York, pp 29–32
62. Cheng Y-T, Cheng C-M (1998a) J Appl Phys 84:1284–1291
63. Cheng Y-T, Cheng C-M (1998b) Appl Phys Lett 73:614–616
64. Cheng Y-T, Cheng C-M (1999) Int J Solids Structures 36:1231–1243
65. Cheng Y-T, Cheng C-M (2004) Mater Sci Eng R44:91–149
66. Zhao J, Frankel GS, McCreery RL (1998) J Electrochem Soc 145:2258–2264
67. Kelly RG, Scully JR, Shoesmith DW, Buchheit RG (2003) Electrochemical techniques in corrosion science and engineering. Marcel Dekker, New York, pp 285–288

The manufacturer's authorised representative in the EU is Springer Nature Customer Service Centre GmbH, Europaplatz 3, 69115 Heidelberg, Germany. If you have any concerns regarding our products, please contact ProductSafety@springernature.com

Printed and bound by CPI Group (UK) Ltd, Croydon, CR0 4YY
30/09/2026
02225558-0006